Essentials of Animal Behaviour

Essentials of Animal Behaviour is an introduction to the study of animal behaviour and is primarily intended for first or second year undergraduates attending short courses in the subject. The book concentrates on putting across the basic principles as briefly and lucidly as possible with the aid of carefully selected examples from both the recent and classic literature, together with numerous illustrations. It will enthuse readers with this active and exciting area of research, and will lay a solid foundation on which further study may be based. Its simple and readable style, helped by an extensive glossary, will also make it useful to senior level school students, their teachers and those with a general interest in the subject. It will be particularly rewarding for all those needing the basics in animal behaviour, behavioural ecology and comparative psychology.

The Institute of Biology aims to advance both the science and practice of biology. Besides providing the general editors for this series, the Institute publishes two journals *Biologist* and the *Journal of Biological Education*, conducts examinations, arranges national and local meetings and represents the views of its members to government and other bodies. The emphasis of the *Studies in Biology* will be on subjects covering major parts of first-year undergraduate courses. We will be publishing new editions of the 'bestsellers' as well as publishing additional new titles.

Titles available in this series

An Introduction to Genetic Engineering, D. S. T. Nicholl

Introductory Photosynthesis, 5th edition, D. O. Hall and K. K. Rao

Introductory Microbiology, J. Heritage, E. G. V. Evans and R. A. Killington

Biotechnology, 3rd edition, J. E. Smith

An Introduction to Parasitology, Bernard E. Matthews

Essentials of Animal Behaviour

P. J. B. Slater
Kennedy Professor of Natural History
University of St Andrews

PUBLISHED BY THE PRESS SYNDICATE OF THE UNIVERSITY OF CAMBRIDGE
The Pitt Building, Trumpington Street, Cambridge CB2 1RP, United Kingdom

CAMBRIDGE UNIVERSITY PRESS
The Edinburgh Building, Cambridge CB2 2RU, UK http://www.cup.cam.ac.uk
40 West 20th Street, New York, NY 10011-4211, USA http://www.cup.org
10 Stamford Road, Oakleigh, Melbourne 3166, Australia

First published 1999

Printed in the United Kingdom at the University Press, Cambridge

Typeset in Monotype Garamond 11/13 in QuarkXPress™ [SE]

A catalogue record for this book is available from the British Library

Library of Congress Cataloguing in Publication data

Slater, P. J. B. (Peter James Bramwell), 1942–
Essentials of animal behaviour / P. J. B. Slater.
p. cm. – (Studies in biology)
Includes bibliographical references and index.
ISBN 0 521 62004 X (hb). – ISBN 0 521 62996 9 (pbk.)
1. Animal behavior. I. Title. II. Series.
QL751.S616 1999
591.5–dc21 98–34540 CIP

ISBN 0 521 62004 X hardback
ISBN 0 521 62996 9 paperback

Contents

Preface

The aim of this book is to provide a brief, but reasonably comprehensive, introduction to the study of animal behaviour. It has grown out of my earlier book, *An Introduction to Ethology*, which was published 12 years ago and has been out of print for some time. Twelve years is a long time in an active field of science, and a lot of interesting things have happened during that period, so the book needed a good going over and sprucing up. But fashions change in science, as in all else, so some parts of the book, describing fields that have seen little recent work, are much as they were before, while I have written new sections and modified the text extensively in those areas in which there has been a lot of recent interest. One obvious change is in the title: the word 'ethology' to describe animal behaviour, and 'ethologist' for someone who studies it, while admirably concise, have rather fallen from use. Unfortunately they have tended to be tied in people's minds to the particular school of study and theories that emanated from Europe in the middle years of this century. Many of these theories have not been supported by later work, and the word ethology has tended to sink with them. It probably sank somewhere in the middle of the Atlantic, as it never really made it to the far side anyway!

The theories of the ethologists were clear and simple and provided a magnificent body of hypotheses on which subsequent work could be based. For this reason, the approach I shall take here is a somewhat historical one, looking at the ideas of the early ethologists, notably Konrad Lorenz and Niko Tinbergen, and then surveying developments over the past few decades to identify subsequent work which has been influential in leading to what we now think about animal behaviour. I have tried to choose some of the best examples, from among both classic studies in the field and the great range of recent work, to illustrate my themes as clearly as possible for those with no previous knowledge of the subject. These have the excitement of approaching this fascinating subject for the first time; if I can share some of the thrill I felt when I did so, this book will have succeeded.

I remain grateful to those who read sections of my earlier book. Three referees made very useful comments and suggestions about what to change and what to leave as it was, and the ideas of various colleagues have helped me with parts of the book, while Jeff Graves read the glossary and Vincent Janik made helpful suggestions on the whole text. It is a particular pleasure to thank Nigel Mann for his delightful illustrations, as well as Jan Parr for some that have been carried over from my previous book, and Tracey Sanderson and Jane Bulleid for seeing the book speedily through the presses.

P. J. B. S. St Andrews, 1998

1

Asking questions about behaviour

People have been fascinated by the behaviour of animals for a long time. Their interest was caught both by the eye-catching activities that they could see around them in the natural world and by the need to understand and control the behaviour of their own domestic animals. Questions naturally arose in their minds. Aristotle, for example, wondered where swallows went in the winter and, seeing them gathering in reedbeds, he speculated (as he did about a lot of things!) that they hibernated in the mud at the bottom of ponds.

But the scientific study of behaviour is a recent phenomenon and, as with so much else in biology, it received its most important boost from the writings of Charles Darwin. Darwin included a chapter in *The Origin of Species* on 'Instinct', a term used in his time to refer to the natural behaviour of animals. He also wrote a book specifically about behaviour called *The Expression of the Emotions in Man and in Animals*. Despite this, in the half century after Darwin there was little work on the behaviour of animals, while zoologists grappled with trying to understand the fundamental principles of systematics, physiology and developmental biology. A few scientists from that time, like Julian Huxley in Britain, Oscar Heinroth in Germany and Charles Otis Whitman in America, stand out for their contribution to behaviour, but they were a small band. It was only in the 1930s that a comprehensive theory of animal behaviour began to emerge through the writings of Konrad Lorenz and, later, of Niko Tinbergen. These men founded the European school of ethology, 'the biological study of behaviour' as Tinbergen defined it. In 1972, this field really came of age as a science when Lorenz and Tinbergen received the Nobel Prize for physiology. They shared it with Karl von Frisch, who discovered the remarkable dance of the honey bee, which enables foragers to tell others in their hive the location of good food sources. This prize was recognition indeed that these three men, whom many saw simply as naturalists, had made a fundamental and lasting contribution to science (Figure 1.1).

It is, in fact, excusable to think of the study of animal behaviour as being a branch of natural history, for the diversity of nature has always been a source of interest and wonderment to those who, like ethologists, studied the natural behaviour of animals. Observing and describing exactly what animals do is fascinating in its own right, and it is also an essential prelude to a more scientific analysis of their behaviour, as was stressed by the early ethologists. It pays to know your animal! Thus spending long hours patiently watching animals (Figure 1.2) can, in itself, be quite revealing even if it is not extended to forming hypotheses and carrying out experiments.

As a result of this preliminary period of thorough and careful description one can make an inventory, or ethogram, of the behaviour patterns of the species being studied. To the casual observer it might look as though different species of birds or of fishes behave much the same as each other. It might also seem that the behaviour of each animal is a highly varied business, not easy to split up into particular categories. Fortunately, for most animals, these impressions are not totally true. Each species tends to have an array of stereotyped behaviour patterns, some of which may be shared with related species but others of which are unique to itself. Describing them and recognising them each time they appear is not as difficult as it might at first seem.

1.1 A case history

Let us illustrate this point with an old favourite of behavioural studies, the three-spined stickleback, an especially easy species to study as it behaves more or less normally in an aquarium tank. Male sticklebacks come into breeding condition in the spring when daylength increases and the streams in which they live become warmer. They change colour, becoming bright red on the underside and iridescent blue on the face, and their behaviour also alters. They gather weed and collect it at a particular spot on the bottom of their pond, gluing it together to form a nest with a special movement that extrudes a sticky secretion from their cloaca. If another male approaches, the territory owner will chase him away or, if he persists, threaten him by adopting a head-down posture which shows off the red belly in all its brilliance (Figure 1.3*a*). Signals such as this are known as displays. If a ripe female stickleback appears, her belly swollen with eggs, our territory owner behaves quite differently, showing another display known as the zig-zag dance (Figure 1.3*b*). He darts alternately

Figure 1.1. Ethology's three Nobel prize winners: Konrad Lorenz, Niko Tinbergen and Karl von Frisch.

Figure 1.2. Though much of the study of animal behaviour today takes place in the controlled environment of the laboratory, the essence of the subject is to understand behaviour as it occurs in nature and, for this reason, much ethological research is still conducted in the wild (photograph © K.J. Stewart).

towards and away from the female in a very striking manner and, if she follows him, he draws her slowly towards his nest. Once there she may creep through the nest and spawn, and he will then follow, fertilising the eggs she has produced as he does so. Her part is then over; indeed he is likely to chase her away, for care of the eggs, and of the young after they hatch, is carried out by the male alone. When he has eggs he stays close by the nest and repairs any damage that it may suffer, as well as showing fanning, a movement that serves to drive a stream of water over the eggs and so keep them supplied with oxygen (Figure 1.3*c*).

This description allows the identification of certain behaviour patterns which are common to all male three-spined sticklebacks in breeding condition: 'gluing', 'head down posture', 'zig-zag dance', 'creeping through', 'fanning'. All these would appear in an ethogram of this species. But the description also raises a great many questions, and it is here that the scientific aspect of ethology begins. Niko Tinbergen recognised that the sorts of questions one could ask about behaviour fell into four different categories: questions about development, about causes, about functions and about evolution. Interest in development might lead one to ask how a male comes to behave in

Figure 1.3. Some of the fixed action patterns shown by a male three-spined stickleback in breeding condition. (*a*) The zig-zag dance with which the male leads the female to the nest, at which he shows her the entrance (*b*) and she creeps through and spawns (*c*); (*d*) shows the male fanning at the nest and (e) the head-down threat display he shows when another male approaches.

the way that he does during the course of his lifetime. For example, does his skill at nest building improve with practice? Does he court a female the very first time that he sees one or must he learn that this is an appropriate way to behave towards her? On the subject of causes, one asks about the mechanisms underlying behaviour, and this concerns both internal states and external stimuli. What is it that signals the male to come into breeding condition in the spring, and how does this affect his physiology so that he is ready to fight and to court? He behaves differently towards females and towards other males: what difference between them leads him to do so? Functional questions concern the advantages of behaving in a particular way. Why does the male show the zig-zag dance rather than simply swimming to the nest? Does his head down posture actually deter other males from approaching, as it should do if its function is to act as a threat signal? Finally, we can ask about the evolution of behaviour. Comparison between different species of sticklebacks can give us clues about how ancient or recent are particular forms of behaviour. Comparison between displays and other behaviour patterns may suggest what actions formed the originals from which displays have been derived.

This account of the questions with which the study of behaviour is concerned makes it sound as though the subject is a straightforward one attacking well-defined problems and without any great controversy. Such an impression would be totally wrong, however. Scientists interested in behaviour have had to contend with gales blowing from various different directions during their short voyage, as well as with awkward questions from some of their own number determined to rock the boat. This has been no bad thing. The early ethologists, and particularly Konrad Lorenz, put forward sweeping general theories based more on careful observation and brilliant intuition than on thorough experimental evidence. These formed a marvellous source of hypotheses for later research, and as a result of this work more detailed knowledge has accumulated and many of the broad and simple theories have had to be replaced. There has also been a growing reluctance to generalise as it has become clear that different animal species vary a great deal in their behaviour so that all-embracing theories are not likely to be very helpful. Rats are not just large mice, far less small people!

1.2 Development and causation

The first real storm to hit animal behaviour blew across the Atlantic, and came in a confrontation between the ethologists and the American school of comparative psychology. The two groups shared an interest in the behaviour of animals, but they approached it from very different viewpoints. The ethologists worked largely in continental Europe and, being zoologists, they had a respect for evolution and were thus interested in a wide variety of species and the different ways in which they behaved. Despite their name, the comparative psychologists at that stage tended not to be concerned with such comparisons and to study very few species, usually just rats and pigeons, their interest being to look for general laws of behaviour that would hold regardless of the species being studied, and preferably apply to humans as well. Their reputation was for rigorous experimental work in carefully controlled laboratory conditions; most ethologists on the other hand simply observed their subjects behaving freely and they did so in the totally uncontrolled conditions of the animal's natural surroundings, those to which selection had adapted it (Figure 1.4). That two such different approaches to very similar topics should lead to

Figure 1.4. A traditional view of the distinction between ethology and psychology was that the psychologist put his animal in a small enclosure and peered in to see what it was doing, while the ethologist put himself in the box and looked out at what the animals round about were up to. Both approaches have their advantages and this distinction between psychology and ethology is now blurred.

confrontation is not surprising. The battle was fought over the subject of behaviour development, a subject on which the views of the two schools were especially starkly contrasted.

The different views on development of the ethologists and the comparative psychologists stemmed, in essence, from the fact that one stressed nature and the other nurture. To most psychologists the learning ability of animals and the flexibility it gives their behaviour is the main interest in studying them, for these aspects may shed light on the equivalent attributes of humans, hence their stress on nurture. To ethologists, on the other hand, the study of species-typical behaviour was a prime concern: they therefore tended to concentrate on patterns which were highly stereotyped and of similar form throughout a species, and they often referred to such acts as 'innate' or 'instinctive'. The assumption here was that nature was all-important and nurture was of little consequence. Indeed, Konrad Lorenz once remarked that the developmental origins of behaviour was a subject of more interest to embryologists than to ethologists.

This controversy was a bitter one, but it was also fruitful, for each side had much to gain from the other, and its resolution brought them closer together. Psychologists came to recognise that evolution had led animal species to be different from each other and placed constraints on what each could learn. For their part, ethologists came to reject the idea that any behaviour was fixed and inflexible and to realise that the acts they studied, no matter how stereotyped, may have been influenced by learning and by other environmental influences. They also came to appreciate the merits of a carefully controlled experimental approach. Thus today many ethologists work in the laboratory and some of them even use the sorts of equipment developed by psychologists for the study of learning, adapted to shed light on ethological questions. In the battle over nature and nurture, the result has been a synthesis: both sides have gained from the realisation that neither nature nor nurture can be ignored in the development of any behaviour pattern. The borderline between ethology and psychology, once hotly contested, has now broken down and those trained in either field may be found working on a variety of topics of interest to both. A good example of work that spans the interests of both fields is given in Box 1.1.

Box 1.1 Development: fear of snakes in rhesus monkeys

Not surprisingly, wild monkeys are normally frightened of snakes. They have special alarm calls that are given when one is about, and these lead others to be very cautious. They will approach and have a good look, but not get too close. Until the snake has passed the behaviour of the whole troop is altered.

Susan Mineka and her colleagues have studied how young rhesus monkeys in captivity come to recognise snakes and be so afraid of them. They have used a piece of equipment (known, rather pretentiously, as the Wisconsin General Test Apparatus), which is simply a plexiglas chamber into which an object is placed. A hungry monkey is then put on one side of this chamber with food on a shelf on the other, and its fear of the object is assessed by how willing it is to reach across. Wild caught monkeys will not do so when there is a snake in the box, but young ones reared in captivity show no such fear, reaching casually across for the food. If, however, they see another monkey being fearful in that situation, even if only on a video screen, they too will become scared next time they themselves are tested. This suggests that the fear is being passed from one animal to another through social learning. But there is more to it than that, as there is something special about snakes that facilitates the learning. If a videotape is edited, so that the tutor monkey appears to be frightened of a bunch of flowers that has been cut-in in place of the snake that was really there, the observer does not become frightened of flowers. So we have an interplay here between learning from others and a predisposition to learn about snakes.

A lot of behaviour development involves such interplays. In this case it is a very neat arrangement. The spectrum of predators in one place may be different from that in another within the range of a species of monkey, and some things that are harmless look very like ones that are not. Learning from the experience of others, but with a bias towards learning to be cautious about some general categories of animal (long thin ones are not a bad start!), is a good developmental strategy, so that the animal comes to avoid things that might harm it but to ignore ones that will not.

Development is just one area that is of interest to both biologists and psychologists studying behaviour. Another is in the field of causation, the study of those outside influences and internal states that lead animals to behave in the way that they do. The senses of animals keep them informed about changes in the external world so that they can react appropriately to many different sorts of stimuli, escaping from those that are potentially dangerous, attempting to capture and eat those that look like food and approaching and courting those that may be prospective mates. Understanding the sensory processes involved is important to shed light on how behaviour is caused, and behavioural studies here border on the interests of the sensory physiologist and the perceptual psychologist. All have the common aim of understanding how events in the outside world are translated into nervous signals and hence into behaviour.

Both physiological psychologists and biologists interested in behaviour may also be interested in how behaviour is influenced by internal events, such as low blood glucose or high levels of a hormone. Just as it is possible to present an animal with a loudspeaker playing a courtship call or a model showing a threat display, so its internal state can be changed: for example, it can be deprived of food so that its blood glucose is lowered or it can be treated with a sex hormone to see whether this affects its behaviour. A full understanding of the causation of behaviour requires knowledge of how both external and internal events affect the nervous system to produce behaviour. Box 1.2 gives an example of a study looking at how these factors interact. To understand causation at all levels also means looking into the exact neural mechanisms involved, the centres and pathways which intervene between senses and movement. This is the realm of the neurobiologist but, from the point of view of understanding the causes of behaviour, it can also be fruitful to treat the animal as a 'black box' rather than probing inside it. For example, one can study the events in the outside world that lead to a behaviour pattern being shown, or how an animal decides which, of the many behaviour patterns in its repertoire, it will carry out at a particular moment. The fact that animals are more willing at some times than at others to show particular responses, like eating, drinking or mating, has led to many different theories of how internal and external factors combine to affect behaviour, collectively known as theories of 'motivation'. Much of the attention of the early ethologists, such as Lorenz and Tinbergen, was devoted to these theories, but recently they have rather fallen from vogue. This is partly because it has become more fashionable to explain the mechanisms underlying behaviour in terms of the animal's neural machinery, as many neurobiologists are trying to do. But it is also because motivation is a very complex matter which requires the taking into account of many different factors for each system of behaviour and cannot be summed up by a simple overall model of the sort that ethologists originally put forward.

Box 1.2 Causation: the incubation of barbary doves

The reproductive behaviour of barbary doves, or ring doves as they are called in the United States, has been studied extensively, particularly initially by Danny Lehrman. If a pair of doves is put together and given a nest and eggs they will not incubate straight away, so the right external stimulus is not all that they need to get the behaviour going. But if a pair has been together for several days, particularly if they have had pine needles with which to build their own nest, they will start to sit when given a nest and eggs even though the female has not laid. Clearly something has happened to change their behaviour.

When the pair are first put together the male starts courting the female, using a display called the 'bowing-coo', with which he struts up and down in front of her cooing (like a dove!). To start with she responds rather little, but after a few days copulation first occurs. At this stage too the male has started carrying nest material to their nest bowl and calling there with another display, the nest-coo. The female follows him to the nest, but leaves when he goes in search of more material. After a day or two, however, she remains behind when he leaves and starts to shape the nest; a further day or two thereafter, when the nest is complete, the first egg appears and they are all set to incubate.

These changes will take place even if the two birds are separated from each other by a wire mesh, so mating itself is not required. But, as far as the female is concerned, the displays of the male are an important factor. Indeed, female doves can be stimulated to lay by the shadow of a male projected onto frosted glass. Interestingly, what these displays appear to do is to encourage the female herself to nest-coo, and this in turn may influence her by self-stimulation. But what leads to incubation? At one level the answer is the stimulation received from displaying and nest building. But what these do is to lead to secretion of hormones. Some of these lead to the production of eggs – obviously also essential for incubation – but the one important for incubation itself is progesterone. If female doves are injected with this hormone before pairing, and given a nest and eggs as soon as they are placed with a male, they will sit at once.

The breeding behaviour of doves is a fine example of how internal factors, such as hormones, and external ones, such as a displaying partner, nest material and eggs, combine to give changes in behaviour. Not only that, but the two birds are nicely synchronised with each other and the changes in behaviour take place at just the right time. Because of their earlier experience, which changed their internal state, the birds are ready to sit as soon as their eggs appear.

1.3 Evolution and function

As mentioned earlier, investigating the evolution of behaviour and its function, or adaptive significance, are two other fields of interest, particularly from a biological perspective. In these areas animal behaviour borders on genetics, evolutionary biology and ecology, rather than on topics of interest to psychologists and neurobiologists. The study of behavioural evolution is not an easy task, for behaviour leaves no fossils, so the best that can be done is to reconstruct the course that it must have taken from a comparison of the behaviour of species alive today. Selection experiments and the study of mutants that behave differently from normal animals can also help us to understand the changes in behaviour that must have taken place during the course of evolution. Rarely, as in the case outlined in Box 1.3, evolutionary change can actually be observed taking place.

The major thrust of recent ethological research has, however, been in the field of function: studies aimed at understanding the adaptive significance of behaviour. The last few decades have seen a revolution in our understanding of evolution theory, with the emergence of many stimulating new ideas. Some of these concepts, such as 'inclusive fitness' and 'evolutionarily stable strategies', have particular relevance to behaviour and have led to intensive study, especially of the social behaviour of animals in the wild, in an effort to discover just how natural selection has led their behaviour to be as it is. This field of study, on the border between animal behaviour, ecology and evolution theory, is usually now referred to as behavioural ecology, and has generated some beautiful work. Some aspects of it, especially the subject of sociobiology, which concentrates on putting social behaviour in an evolutionary framework, have also led to a great deal of controversy, much of it rather fruitless. One area of argument has been over the relevance of these ideas to humans, as some sociobiologists (often now calling themselves evolutionary psychologists, just to confuse!) are enthusiastic about the application of evolutionary ideas to humans, believing that our behaviour can be best understood if viewed in the context of our evolutionary heritage.

BOX 1.3 Evolution: the behaviour of guppies in Trinidad

Guppies are small fish that occur in the streams and rivers of northeastern South America and a few Caribbean islands, and they are also common aquarium fish as they are easily kept and bred in captivity. The males have attractive iridescent patches of various colours with which they display to the females. They have been studied particularly in Trinidad by a succession of people, most recently Anne Houde and John Endler, and Anne Magurran and Ben Seghers together with various collaborators. In Trinidad guppies occur in numerous rivers, though obstacles such as waterfalls have sometimes halted their spread. Their morphology and behaviour varies between the different rivers and this has caused particular interest: it provides a rare opportunity to study evolution in action.

A critical factor turns out to be the distribution of predators between the rivers. Some sites are 'high risk' with a range of predatory fish, such as pike-cichlids, while

others are 'low risk' with rather few of them. At the high risk sites males are less brightly coloured and there is some evidence that the females in these areas may prefer them that way, unlike their usual preference for the brightest. In places where a predatory shrimp is common, females prefer males with orange colouration; orange is a colour the shrimp cannot see.

When it comes to behaviour, fish in the high risk sites spend much more of their time in schools, and they are more tightly grouped within these. They are also more cautious when a predator is present, keeping a greater distance from it than fish from low risk areas. That such differences have a genetic basis has been confirmed by the fact that the differences in schooling behaviour persist after several generations of breeding in captivity. Further evidence comes from an experiment where 200 guppies were moved from a high risk area to a virtually predator-free one, above a waterfall where there were previously no guppies. Captive breeding from this population 34 years later showed that their offspring schooled less and were not as wary of potential predators as their ancestors at the high risk sites. Schooling has costs, such as increased competition for food, so that selection might be expected to lead to a decrease in the absence of predators. It is striking that such a decrease could be found after only 100 generations or so.

Guppies breed rapidly and have a short generation time. Not many other species give us opportunities to observe evolution actually happening in the wild. However, as we will see in Chapter 6, comparisons between species can often allow us to reconstruct the sorts of changes that must have taken place during their evolution.

Another, related, bone of contention has been whether or not the claim that behaviour is adaptive means that it is unmodifiable and under tight genetic control: some sociobiologists have written as if it did mean this, and have hence been charged with 'genetic determinism', an especially heinous crime if they are also writing about humans! But, in truth, all that is required for behaviour to be acted on by natural selection is that it has some genetic basis, no matter how slight. There is no reason why its development should be in any sense tightly controlled, so adaptation does not presuppose genetic determinism. And, in the case of humans, behaviour may come to be adaptive for a great many reasons other than through natural selection, notably through our remarkable ability to learn by our own experience and from the experience of others. Indeed, for many of us the environment we occupy is so odd compared with that in which we evolved that it would be hard to argue that our adaptation to it had much to do with selection! Yet we have both modified our environment to suit ourselves and adapted ourselves to it in a fine degree of detail.

These controversies have generated a good deal of heat, and there are still those who are fiercely committed to one side or the other without a hint of compromise. But reason, as so often, takes a middle course, and neither extreme position is very plausible. Both evolutionary heritage and the genes which are its legacy from one generation to the next clearly influence all the behaviour of all animals, ourselves included, but to argue that either is all-important is as mistaken as to deny that either has any importance. A more serious difficulty with functional reasoning, as with much evolutionary thought, is the ease with which hypotheses can be generated but the difficulty of subjecting them to any rigorous test. Writing popular books, packed with brilliant hypotheses dressed up as facts about the relevance of animal behaviour to humans, is quite a money spinner. Many of the 'just so stories' with which these are filled are so compelling that one feels they just *must* be true – until someone else comes up with an even neater explanation!

The problem is that functional questions are not often easy to test experimentally. To carry out an experiment what one needs is two groups of animals, an experimental and a control, which differ only in that aspect of their behaviour one wants to study. It is not easy to achieve such a very specific change, leaving all other aspects of behaviour the same. Furthermore, adaptation is only relevant to the particular environment in which the species evolved, so the study is most likely to be useful if it is carried out in nature, which further limits the sorts of experiments that can be done. Faced with such problems, many behavioural ecologists have rejected experimentation and concentrated on observation and correlation instead. The results of this can be persuasive, particularly, as in the example shown in Box 1.4, where they are geared to testing specific hypotheses, in this case that bumblebees forage in the most efficient possible manner.

But a correlational approach has problems of its own. The size of monkey troops may correlate with the size of their home ranges, but this fact does not, in itself, provide an explanation. A larger group may require a larger range to feed on, a larger range may require a larger group to defend it, or both features may be caused by a third factor and not directly related to each other at all. For instance, a food supply which is briefly abundant in a small area, then dies out and 'blossoms' once more some distance away, may lead to both. It may encourage a large home range so that there is always at least one source of food present in it. Furthermore, as each source in turn is abundant, group size can be large without disadvantage to the members of the group. Thus large group and large range will both arise without one causing the other. The moral is that one must be cautious about interpreting correlations and that, in function as in other aspects of behaviour, greater certainty can be reached if it is possible to carry out experiments. This is not easy but, with ingenuity, some clever tests of functional hypotheses have been devised.

BOX 1.4 Function: the foraging behaviour of bumblebees

The relationship between flowers and the insects that pollinate them is a close and fascinating one. Insects fly from flower to flower feeding on the nectar that is produced for them as a reward for carrying pollen from one flower to the next. But this is only of benefit to the flower if the insect takes the pollen to one of its own species, for pollen is the plant's equivalent of sperm and is wasted unless it fertilises an egg of the same species. Insects, however, do tend to specialise on the flowers of one species because skills are usually necessary to reach the nectar and, having developed these, they are better equipped to do so. Indeed flower species probably differ from each other in shape very substantially for this very reason: only insects that have developed the appropriate skills can feed on them, and this forces the individual insect to stick to the same species and thus carry pollen from one plant to another of the same sort.

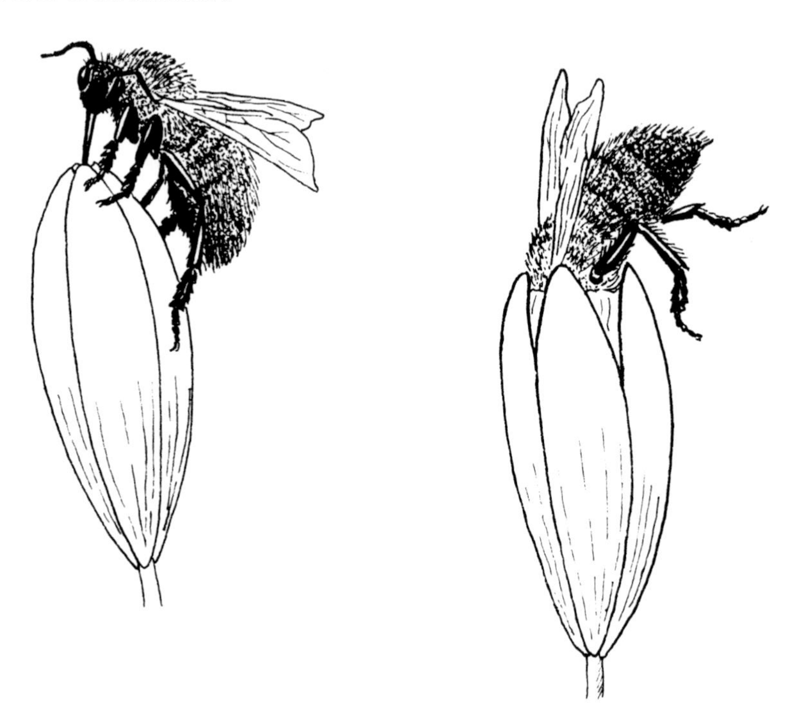

Bumblebees are no exception to this rule, and the adaptiveness of their behaviour has been shown beautifully by Bernd Heinrich. Each bee specialises on one particular flower, referred to as its 'major', while taking occasional nectar meals from other species, its 'minors'. Worker bumblebees are short lived and tend to stick to the same major, but queens live longer and may switch from one to another. Having minors probably enables them to check up on whether another flower has become more profitable than their current favourite. As flowering seasons are short, the abundance of different sorts of flower changes through the year and thus bees are well advised to check that their major still gives the best returns.

Whether it is worth a bee feeding on a particular flower species depends on several factors. Some flowers are much more common than others, so there is little flying time between one meal and the next. Some flowers also produce much more nectar than others so that the meals they provide are larger. Finally, the profitability of a type of flower depends on how many bees are feeding on it. If lots are doing so, then a bee looking for a food source will find that flower rather unprofitable, for there will be little nectar available on each visit. As a result, it is only common flowers with a high rate of nectar production that have large numbers of bees feeding on them. Each bee arrives shortly after another one and so receives little reward, but then it does not have to fly far to the next meal. On the other hand, few bees specialise on rare flowers with a low nectar yield and so each of them gets a large meal at each visit, though it has to fly a long way between one flower and the next to get this reward. Taking all these points into account, Heinrich argued that every bee does about as well as every other in terms of nectar intake per unit time. It follows that it would not pay any of them to switch to a different flower from the one they are majoring on except if these change in frequency. Their behaviour is thus beautifully adaptive to themselves, as well as to the flowers.

1.4 Keeping things simple

One nice thing about animal behaviour as a field is that it is relatively free of jargon compared with many other branches of biology. A glance at the glossary at the end of this book will yield few words that are totally unfamiliar. But there is a hidden problem here, for many of these words are used in a special sense that is not the same as their everyday meaning, and is usually more precise and more restricted. The word function is a good example here. While it is in the vocabulary of most people, rather few mean selective advantage when they say it! There has also been a recent tendency to adopt words from everyday language, which are not entirely appropriate when applied to animal behaviour. Male animals may often mate with unwilling females, and for a while the word rape was used to describe this. But that word carries with it a good many overtones, for example of culpability, which one would not want to apply in animal examples. To refer to 'enforced copulation' may be more cumbersome, but it is also safer. The same argument could be applied to words like 'lying' and 'cheating', though these are now firmly in the behavioural vocabulary. It is important to realise that these words do not imply any malice or evil intention on the part of animals, or indeed that they are thinking anything out for themselves at all. A bird may avoid eating a brightly coloured butterfly that tastes nice because its colour pattern mimics that of another species that tastes nasty. The bird is certainly deceived, but to label such a beautiful adaptation as 'lying' on the part of the butterfly is really going too far!

Problems in the use of words like these are part of the more general issue of 'anthropomorphism', looking at animals as if they were people. The way in which they behave often looks extraordinarily clever and sophisticated, as if they had thought about the problem and decided what the best thing to do was. But beware! Millions of years of evolution can come up with an answer, often in very simple animals with small nervous systems, based on a few rules, which looks not dissimilar to the solution to the same problem we might work out for ourselves. Animals may indeed think things out, have intentions, and behave in all sorts of other ways like we do, but these are very difficult ideas to test scientifically. One of the most important rules in science is that known as 'Occam's razor' which states, in essence, that one should test the simplest hypothesis first and only if it is found wanting move on to more complex ones which are less easy to disprove. Fortunately, as we shall see, some delightfully simple ideas have proved adequate to account for many aspects of the behaviour of animals.

Related to the assumption that animals have intentions is to write as if natural selection had. This is the problem of 'teleology': the implication that there is purpose or design in nature. Natural selection does not act 'in order to' achieve something, but works in retrospect, successful traits persisting and unsuccessful ones dying out. It is a common shorthand among biologists, nevertheless, to write as if selection had purpose; as with the actions of animals, it is best not to for fear of misunderstanding.

Having come up with these strictures, I had better try to abide by them myself without resorting to contorted sentences and leaden prose!

1.5 Book plan

The study of animal behaviour concentrates on four different types of question: causation, development, evolution and function. These topics will form the core of this book and are dealt with, each in turn, in Chapters 4–7. But, as the best studies have always started with a period of description, we will follow this tradition in the next two chapters by describing the motor patterns of animals, their form and what is known of their control, then the senses which give the animal its outlook onto its environment, and how stimuli influence these to produce behaviour. Much of the book will be concerned with individual animals, but we will often refer to their interactions with one another, as when they fight or they mate. In the last two chapters of the book we will come to the social aspects of behaviour more specifically, to discuss communication and the social organisations which are built up by the communicative interactions between individual animals. At the end of the book there is a glossary. This deals with words and phrases in common use among those studying behaviour. Most of these will be found elsewhere in the book, and referring to the glossary may help to remind the reader of their meaning, but I have included others as well in the hope that this dictionary will help the reader who takes the subject on through wider reading. The literature list at the end of the book will, I hope, also help in this. It is not intended to be exhaustive, but gives a list of books in which the reader will be able to take the topics of each chapter to greater depth, and journals where it is worth looking for the latest findings.

2

Patterns of movement

The ways in which we define the behaviour patterns of animals depend, almost entirely, on the exact movements involved. These movements vary enormously in their complexity, from the slow sweep of a chameleon's eye, which involves just a few very specific muscles, to the headlong dash of a cheetah, in which most of the muscles in its body have some role to play. The simplest of movements, usually referred to as reflexes, may just involve a few sensory cells connected through two or three nerve cells to a few muscle fibres, so that they make a good starting point in thinking about behaviour.

2.1 Reflexes

We all know about several of our own reflexes: the blink of the eyelids which occurs involuntarily when something flies towards one's face, the constriction of the iris caused by the flash of a bright light, the knee-jerk reflex induced by a tap just beneath the knee cap. All these require a stimulus from the outside world, but it is a simple one and they are otherwise quite automatic. The knee-jerk, for instance, involves only two nerve cells (Figure 2.1). The connection between these is far down in the spinal cord (though admittedly that is quite a long way *up* from the knee), but the point is that the brain need know nothing of the action, for it is all carried out involuntarily and does not require any complex processing. This is even so where a third nerve cell, or interneurone, occurs between the sensory and motor nerves, as often happens in reflexes: the interneurone is short and all the connections are far away from the brain itself.

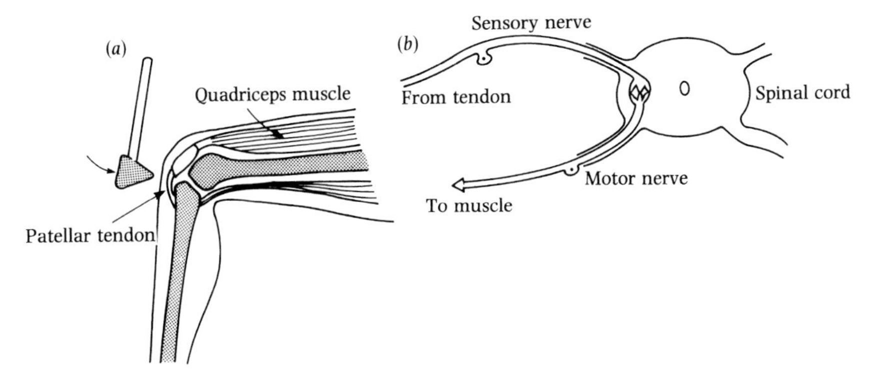

Figure 2.1. The knee-jerk reflex depends on sense organs in the patellar tendon just below the knee cap and on the quadriceps muscle in the upper leg (*a*). Though close, these are not simply connected to one another. The sensory nerve travels up to the spinal cord in the rump region and there synapses with the motor nerve which passes back down the leg to the muscle (*b*). As it involves only two nerve cells, this is one of the simplest of reflexes.

One of the first people to study reflexes was Sir Charles Sherrington, the great English physiologist, who examined various reflexes of dogs in the early years of this century. For much of his work he concentrated on the 'scratch reflex', which dogs use to remove irritation from their flanks with their hind paws, and he studied this using the simplest of apparatus. The dog's leg was attached by a thread to a needle which drew a trace on a rotating smoked drum. If the paw was still, the trace was made horizontally as the drum went round but, if the dog scratched, the needle made a line up and down on the drum as it did so (Figure 2.2).

Many of the properties of reflexes Sherrington found also apply to more complex actions, but the fact that they occur even in the simplest acts points to them being fundamental features of behaviour. Two of the most obvious of these characteristics are latency and after-discharge: when the dog's flank is touched there is a short delay before scratching starts and, when the stimulus is removed, again there is a slight gap before the dog takes away its paw. These are simply explained: it takes time for the nervous signal to travel from the touch receptors in the skin to the muscles responsible for scratching. Signals travel very fast along nerves, but there is a greater delay where they must cross from one nerve cell to the next within the spinal cord. As a result the dog is always a little out of date in its behaviour and scratching does not start for a short while after the irritation arrives nor cease till some time after it has gone.

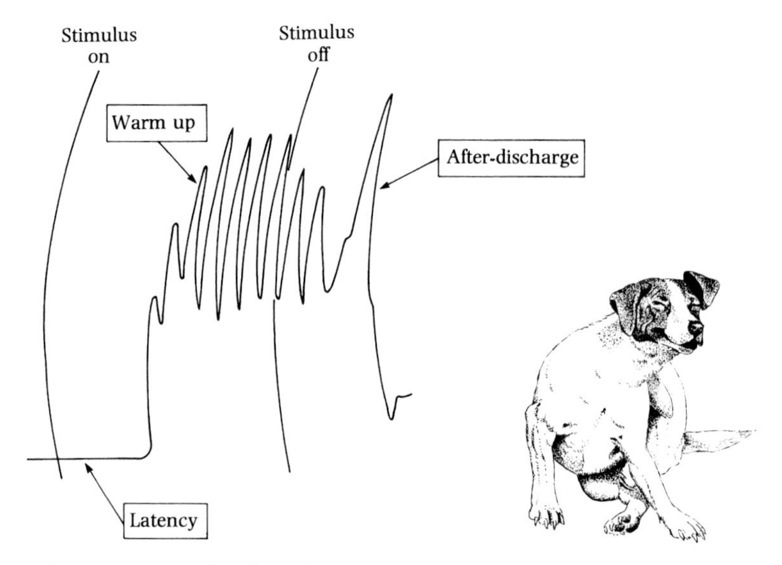

Figure 2.2. The scratch reflex of a dog as studied by Sherrington. The sample trace from one of his experiments shown here illustrates three features of reflexes that he demonstrated: latency, warm-up and after discharge. (After Sherrington, C.S. 1906. *The Integrative Action of the Nervous System*. Cambridge University Press, London.)

Two less obviously explained properties of reflexes are 'warm up' and 'fatigue', but these two are also important and have their counterparts in other aspects of behaviour. Warm up refers to the fact that the behaviour, once started, does not reach maximum intensity straight away. In the case of the scratch reflex, the first few strokes with the paw do not have such a broad sweep as later ones, it taking a little time for the action to reach full amplitude, probably as more and more muscle fibres are recruited. More complex behaviour patterns change in an equivalent way. The rate of alarm calling of a small bird that has spotted a hawk rises over a minute or two rather than being at a maximum as soon as the hawk appears, and an eating mouse nibbles slowly at first and then faster and faster as its meal progresses.

Fatigue is also a common occurrence, in reflexes as in other behaviour. In the case of the dog, it can be seen quite simply by touching the animal's side with something which cannot be removed by scratching. The dog will start to scratch and, after a period of warm-up, the intensity of scratching will reach a peak, but it will not persist indefinitely. Slowly it dies away until the animal scratches no more despite the fact that the stimulus is just the same as it was before it began. Something in the system is clearly fatigued, and needs a little time to recover before scratching can start again.

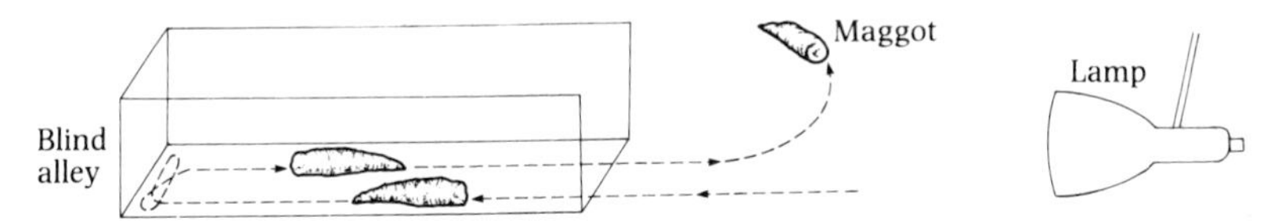

Figure 2.3. *Sarcophaga* larvae move away from light but, if this leads them into a blind alley, they escape from it after a few minutes and then carry on the way they were going. This highly adaptive response results simply because their ability to turn becomes exhausted in the alley and all they can do is move out of it. Though it may look intelligent, their behaviour results from some simple properties of their nervous system.

Fatigue is just one aspect of changing motivation: the fact that animals are not always equally prepared to show a particular behaviour pattern when the stimulus appropriate to it appears. In the case of fatigue this is for the specific reason that the behaviour has just been elicited a lot, but there are many other reasons why animals may fail to respond and this wider topic is one to which we shall return in Chapter 4. In the case of reflexes, however, the willingness of animals to perform usually varies rather little provided that they are not stimulated too often. If they are, fatigue does set in, and there are various reasons why it may do so. An obvious one might be that the muscles involved in the action are simply exhausted and incapable of any further exertion. Another possibility is that the sense organs are tired out and are no longer capable of detecting the stimulus. The dog may stop scratching because it does not feel the irritation any more.

These are the simplest suggestions to account for fatigue, but it is often possible to discount both of them because the animal's senses and its muscles are used perfectly normally in other activities while it remains incapable of showing the action which has fatigued. The reaction of a fly maggot to light, studied by Mario Zanforlin, is a good example. After they have finished feeding, these larvae search restlessly for a dark and enclosed space in which to pupate safely. If a light is shone onto them from one direction they will turn and move away, but what do they do if they then find themselves in a blind alley? Amazingly, as Figure 2.3 shows, after searching around for a while at the end of it, they move straight out of it again towards the light and then turn away once more when they emerge: a striking example of apparently intelligent behaviour by an animal no one would accuse of brilliance! Could it be sensory adaptation? Perhaps the animal ceases to see the light when in the alley but turns away on emergence because the light is much brighter at the mouth than at the far end. This is not the answer: the maggot will react in the same way even if the light is slowly withdrawn as it emerges so that its intensity remains the same. Furthermore, maggots free to move away from light indefinitely will do so for several kilometres without once moving towards it. Nor is the result simply due to muscular fatigue, as the animal remains fully capable of all its other reactions using the same muscles. Instead the larva is experiencing a very specific deficit: it is incapable of making sharp turns. In a narrow alley it must make these if it is to stay at the end, but after a few of them its turning reaction becomes exhausted and thus the only direction in which it can move is straight out of the alley. It then encounters a wide open space in which it can make a broad sweeping turn to take it away from the light again. But it will be some minutes before its nervous system recovers enough to allow it to make any more of those tight turns that it first made when it reached the end of the alley. The evidence is against its fatigue being sensory or muscular, but the incapacity probably arises because certain nervous pathways become unable to carry signals, perhaps through depletion of the transmitter on which they depend for contact between one cell and the next. The pathways need a rest before they can be used again, and during that time the animal cannot show actions that depend upon them.

2.2 More complex behaviour patterns

Although illustrated with simple reflex actions, these points are equally applicable to more complex behaviour patterns, such as the courtship and threat postures of sticklebacks referred to in the last chapter. Actions such as these were called 'Fixed Action Patterns' (or FAPs) by ethologists, but this term has fallen from use recently as they have been studied more and their fixity called into question. There is no hard and fast line between such actions and reflexes, but they do tend to be rather more complicated and to vary in whether or not they can be elicited regardless of whether they have been subjected to 'fatigue'. Many of them are only produced in response to quite complicated stimuli: for example, size, shape, hue and texture may all contribute to the ability of a visual stimulus to effect a response. This is a topic to which we shall return in the next chapter: it is enough to point out here that this is a far cry from the sharp tap or the change in light intensity which may be the critical stimulus for a reflex. Integrating such complex information from the outside world obviously also involves more than a few neurones: processing in the sensory areas of the central nervous system may be crucial in the decision of whether or not action is taken. And the action itself, no matter how stereotyped, will often involve hundreds of different muscles all contracted in well ordered sequences, again involving many neurones.

Figure 2.4. The head-toss display of the goldeneye drake. Like many duck displays this is very stereotyped, the male suddenly tossing his head back and then bringing it forward again in almost exactly the same way every time.

The idea that animals possessed a repertoire of Fixed Action Patterns originated with Konrad Lorenz. He described the properties of FAPs very clearly and this makes a useful starting point in considering how ideas have changed in the intervening half century. One of his claims was that these acts are innate, and indeed the original German word he applied to them was 'Erbkoordination', meaning literally 'inherited coordination'. But, as pointed out in the last chapter and discussed in more detail in Chapter 5, this is not a very useful criterion as all behaviour depends on an animal's inheritance so no sharp line can be drawn between actions that are innate and those that are not. Another suggestion was that fixed action patterns are 'invariant', showing negligible differences between different individuals or between repetitions by the same individual. Indeed this is implied by the word 'Fixed'. But detailed analysis shows that this is not as true as it seems at first sight. Some actions are certainly very stereotyped, and this is especially true of many courtship signals for the simple reason that one of their main functions is to indicate to prospective mates the species to which the individual belongs. A well studied example here is the head-toss of the male goldeneye duck (Figure 2.4). Woe betide the bird whose head-toss diverges from the norm, for he will get no mate. The average head-toss is 1.29 seconds long, and the standard deviation, which is a conventional measure of variability, is only 0.08 seconds. What this means is that 95% of tosses are between 1.13 and 1.45 seconds in length, which is indeed pretty constant. But not all actions are as fixed as this and it would be artificial to draw a line, calling some of them FAPs and some not. Rather than doing so, the term has tended to drop from use and most of those

Figure 2.5. Like most mammals, horses walk by taking their diagonally opposite legs off the ground at the same time. Giraffes, by contrast, lift both their left legs and then both their right ones

studying behaviour today simply refer to 'behaviour patterns' or 'acts'. Nevertheless, it is true that many animal species do have an array of stereotyped and species-typical patterns of behaviour. Some may only be shown by males in breeding condition, some only by infants, some by lactating mothers, but within each category all animals will show them in more or less the same form.

Some contrasts between species highlight the way in which behaviour patterns are the same throughout a species but may be different between them. When they walk, most mammals do so with their diagonally opposite legs off the ground at the same time, which undoubtedly helps to keep their centre of gravity above the two that are in contact. But, perhaps surprisingly for an animal with its centre of gravity quite so high, the giraffe does not (Figure 2.5). It raises both left legs, then both right ones, swinging its body from side to side as it goes. This falls neatly into the category of a stereotyped and species-typical aspect of behaviour: all horses walk one way, all giraffes the other, though horses can be trained to 'pace' following the giraffe pattern. Drinking in birds provides another contrast. Most birds cannot swallow water with their heads down but take a sip and then raise their heads to let the water slide down their throats. Pigeons, on the other hand, do not have this problem but simply suck the water down. Again this is a hard and fast rule: one will not find the odd sparrow that has mastered the art – nor the odd pigeon that has not. These distinctions are sometimes useful to taxonomists, in their efforts to classify species of animal into groups, for behaviour can be as much a species-typical characteristic as can number of toes or coat colour. Among wading birds, one of the reasons why the oystercatcher is classified as closely related to the plovers depends on its behaviour. Like a plover it scratches its head by raising the leg over the top of the wing, whereas other waders bring the leg up under the wing.

These examples illustrate the fact that closely related species may often share patterns of behaviour with each other. Similarities between species are most striking in the sorts of actions just described, such as walking, drinking or head scratching. They are less often true of courtship acts for, just as the male goldeneye must get it right to ensure he is accepted by a female of his species, his courtship must be sufficiently different from that of other species of duck to avoid attracting their females. To return to sticklebacks for a moment, the three-spined and the ten-spined both show zig-zag dance in their courtship, but the female does not have to resort to counting spines to decide which to mate with! The ten-spined is jet black where the three-spined is red, and his dance is carried out zig-zagging vertically with his head down towards the mud, as if he was hopping along on a pogo stick, rather than to and fro horizontally. Thus, even at a distance, females will have no difficulty in selecting a mate from the right species.

Rather than stressing the innateness and invariance of behaviour, today we tend to think of each species as having an array of relatively stereotyped behaviour patterns which will be shown by all its members, at least of the same age and sex. Subsequent studies have shown that few, if any, behaviour patterns would fit in with the defining characteristics that Lorenz laid down when he first described fixed action patterns. One of these illustrates this rather well: his suggestion that these actions were triggered by stimuli in the external world but, once started, became independent of them. In proposing this, he no doubt had in mind the egg-rolling behaviour of the greylag goose, which he and Tinbergen studied at around the time he first described fixed action patterns.

If an egg rolls from its nest, the goose leans out, places its beak beyond the egg, and then carefully draws the beak back towards its chest so that the egg is retrieved (Figure 2.6). Lorenz referred to this action as having a fixed component, the movement of the beak towards the chest, and a variable component, side to side movements of the beak which serve to keep the egg in place. The latter are modified by the wobbling of the egg but the chestwards movement, Lorenz argued, is independent of the egg once started and so constitutes the truly fixed part of the action pattern. As a demonstration of this, he and Tinbergen removed eggs from geese which were in the middle of rolling

Figure 2.6. The egg-rolling response of a greylag goose, shown when an egg has rolled from its nest. (after Lorenz, K.Z. & Tinbergen, N., 1939. *Z. Tierpsychol.* **2**, 1–29.)

them back: a somewhat risky procedure, as anyone who has attempted to approach an incubating goose will realise! Remarkably, the animals continued with the action till their beaks returned to the nest, even though the original stimulus had been removed. The action thus appeared to be triggered by the egg but not to depend on its presence thereafter.

There are not many other examples of behaviour patterns which run to completion regardless of information from the outside word. Most of them, as we shall see in Chapter 4, are strongly dependent on feedback from what they achieve. The way in which animals track moving objects, such as potential mates or prey, will highlight the difference between behaviour patterns which rely on such feedback and those that do not. If a dog sees a rabbit, it runs towards it watching it all the while, so that if the rabbit moves off it can change its trajectory to maximise the chances that their paths will meet. Generally speaking dogs do not arrive at the spot where a rabbit originally sat and wonder where it went to! Their movements are continually changed in the light of incoming information or, in more general terms, in the light of feedback from the results of their actions. Indeed, if the behaviour of hoverfly males chasing prospective mates is anything to go by, the dog may well do one stage better than this. The hoverflies do not simply aim at where the object they are tracking is now, but they plot an interception course, flying in a direction which takes account of the target's movement and minimises the time it will take to reach it (Figure 2.7).

By contrast with these examples, a preying mantis, if it finds a fly within striking distance, will rock to and fro and then suddenly pounce with

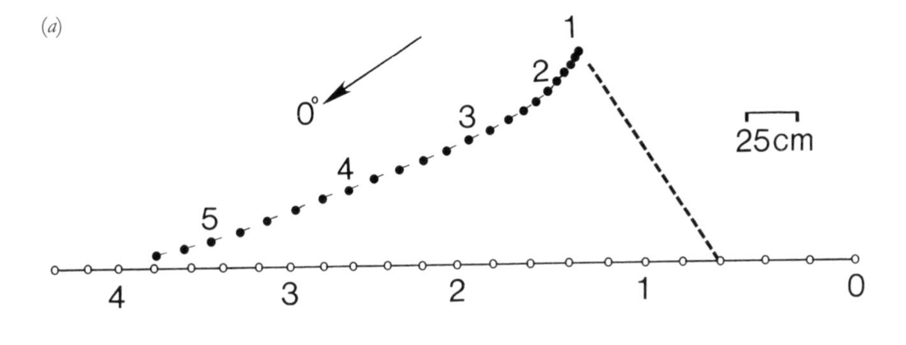

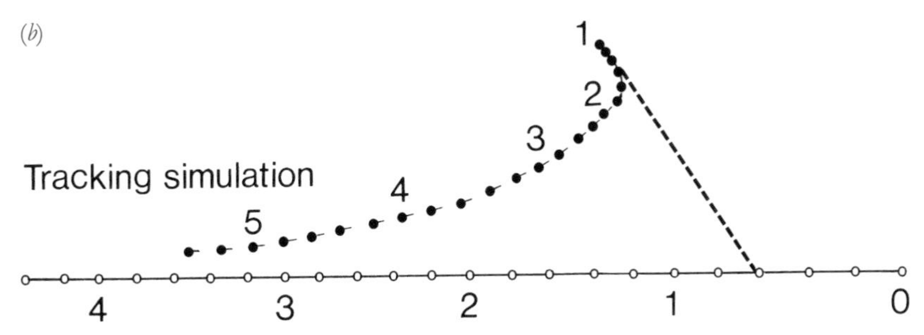

Figure 2.7. (*a*) A male of the hoverfly *Volucella* (black dots) tracking an object fired from a pea shooter (white dots) as it moves from right to left. The position of the fly and its quarry is shown every 20 ms and numbered every 100 ms. The dashed line shows the line of sight between male and quarry before the fly accelerates to intercept it. At each point thereafter the fly is heading rather ahead of the object so as to enhance its chances of interception. (*b*) Shows the route the fly would have taken had it headed straight for the object. (After Collett, T.S. & Land, M.F., 1978. *J. comp. Physiol.* **125**, 191–204.)

alarming rapidity. Should the fly move after the mantis has initiated the pounce, its meal is lost, for it cannot change the course of its strike once it has started. Like the goose, this is an example of an action which fits in with Lorenz's description: it is only triggered by the external world but not modified by it as it proceeds, so that once initiated it goes to completion regardless. In the case of the mantis, it is easy to see why: the action is so rapid that there simply is not time for feedback – nor indeed is there much time for the fly to move beyond the area of its strike. But most behaviour is less rushed than this, and is modified by feedback from its consequences, so that few acts would be defined as fixed action patterns according to Lorenz's strict definition.

2.3 Central and peripheral control

A question related to whether animals sense changes in the external world and modify their behaviour in line with them is whether internal changes brought about by their actions affect how they behave. To some extent this must obviously be so. The more one eats the more satiated one becomes and clearly satiation results from sensing that one has had enough: such factors as having a full stomach and high levels of nutrients in the blood contribute to this feeling. But consider a shorter term action like the sequence of leg movements in walking. This could be organised in two quite distinct ways, one relying on peripheral information and the other not. The brain might send out detailed instructions to each of the relevant muscle blocks in turn, patterned in the right sequence to bring about the relevant leg movements. This would not need any knowledge of what the legs were actually doing. At the other extreme, the brain might simply send a message to the first muscle to contract, this muscle might be linked to the next, and so on until the last muscle in the sequence was linked back to the first again. With such a system, cycles of walking would occur because of the connections from one muscle to the next, the only instructions from the brain itself being the orders to start and to stop.

These are two extreme ways in which movements might be arranged, the one where patterning was central, the other where it lay in connections at the periphery. In practice, motor organisation tends to rely to some extent on both sorts of influence. The sequence of muscular contractions in walking does appear to depend upon central instructions, but its details can be modified by information from sense organs at the periphery. An example of such a peripheral influence is the crossed-extensor reflex which has the effect, when one leg is raised from the ground, of causing contraction in the extensor muscles of its neighbour, so leading it to stiffen and bear the weight of the body. Such actions will obviously help in the adjustments necessary during movements. Similarly, information from the limbs to indicate exactly when they have hit the ground, how much resistance they are receiving from it and how much weight they are bearing, will be essential to control the fine timing of the sequence of actions. Imagine a cross-country runner moving over rough terrain (Figure 2.8). In such a situation a robot from a science fiction film simply following the instructions from its computer without any feedback from what its actions achieved would rapidly fall over. It is no coincidence that such robots are always filmed moving behind some rubble so that one cannot see the smooth path that has been specially laid for them!

These points illustrate the importance of peripheral modulation to adjust movements in keeping with environmental changes. But modulation is, for

Figure 2.8. Feedback enables a cross-country runner to move over even the roughest terrain, modifying the tension in different muscles and shifting his balance to suit the ground he encounters (© Prestige Photography). To achieve similar feats, robots would have to be very complex: most can only cope with smooth paths as they have no feedback system to enable them to respond to the lie of the land.

most movements, as much as the periphery provides, the basic pattern being passed down from the brain. This enables changes of speed and of gait to be imposed from above and it also enables the same muscles to be used for many different actions. If a stimulus to one muscle set in train a series of commands to others so that the animal started walking, the same muscle could not be used for kicking or for scratching, at least without complex connections to inhibit the unwanted sequences which would otherwise follow automatically from its activity. The combination of central patterning and fine feedback control from the periphery assures the utmost precision of movement; one need only watch an Olympic gymnast to be impressed by what such an arrangement can achieve.

The central origin of motor patterning has been most clearly shown in some invertebrate systems. An especially good example comes from silkmoth eclosion: the movements shown by the moth as it twists and turns to free itself and emerge from its pupa (Figure 2.9). This is not what one might at first sight think of as a single behaviour pattern, as the whole procedure takes about one and a half hours to complete, but it is stereotyped and species-typical, and its

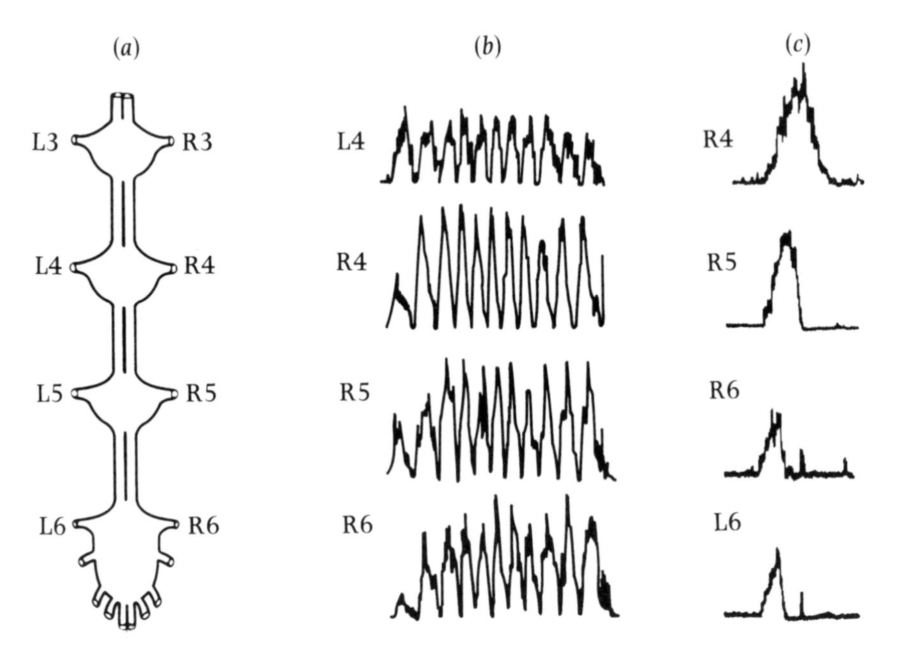

Figure 2.9. Recording from the motor nerves in the isolated abdominal nervous system (*a*) of a silkmoth after eclosion hormone has been added. To begin with (*b*) all the nerves on the left fire alternately with all those on the right. In the traces shown R4, R5 and R6 are firing simultaneously, but in alternation with L4. Then, after a period of quiescence, waves of firing pass up the body (*c*), from R6 to R5 to R4, and now the motor nerves in the same segment fire simultaneously rather than in alternation. Thus L6 and R6 show peaks at the same time in the example shown. (After Truman, J.W. & Weeks, J.C., 1983. *Symp. Soc. exp. Biol.* **37**, 223–241.)

length just highlights how varied behaviour is. Such a long drawn out procedure also has the advantage of being leisured enough to be studied without slow motion equipment! The sequence is split into three half-hour periods. In the first (Figure 2.9*b*), all the muscles down one side of the body contract alternately with those down the other, so that the abdomen twitches from side to side and becomes detached from the pupal case. In the next half-hour the animal rests quietly and in the last it performs a quite different sequence of muscular movements which lead it to emerge from the pupal case. During this period waves of muscular contraction pass from segment to segment down the body, the muscles on the two sides of a segment now contracting at the same time rather than alternately (Figure 2.9*c*). These peristaltic movements, like those of an earthworm as it burrows, have the effect of moving the case off the end of the body and so freeing the moth.

All of these very precise movements are patterned entirely within the four abdominal ganglia of the moth and require no sensory connections whatsoever. If these ganglia are carefully dissected out from the moth, retaining their connections with each other but cutting the motor and sensory nerves leaving them, recordings can be made from the remains of the motor nerves to see what the ganglia on their own can produce. Amazingly, the same sequence is found as in the intact moth. For about half an hour all the nerves on one side fire in alternation with those on the other, then there is a period of quiescence, and then the nerves to both sides fire from each ganglion in turn. The pattern is just the same as normal, but there are no muscles to translate it into action. What then stimulates this sequence if sensory nerves play no role and nor do connections from the ganglia further forward in the body? A trick is needed to get the program started: the sequence will not get under way unless the appropriate 'eclosion hormone' is added to the medium in which the ganglia are kept. This hormone normally starts to circulate in the body when eclosion is due and sets the whole process in motion.

2.4 Orientation of movements

The eclosion of silkmoths is a remarkable example of a behaviour sequence which is complex and yet entirely organised within the nervous system without the need either for sensory influences from outside the animal to trigger it or for sensory feedback to affect its course. This is very unusual, for in most behaviour sensory information has a crucial part to play, even if only to ensure that the movements of the animal are correctly orientated with respect to the outside world. Here an even longer term series of actions, the migration of European warblers, serves as a good example. These small birds fly south in the autumn, some species only going as far as the Mediterranean, but others travelling on over the Sahara into the southern half of Africa. Sensing the correct direction to take depends on outside cues, such as the stars and the earth's magnetic field, but other aspects of this migratory pattern occur regardless of any input from the outside. Even when kept in laboratory cages with constant daylength the birds become restless at night twice a year in spring and in autumn at the times when they would normally migrate. The species which migrate furthest are restless for more nights than those which travel a shorter distance, and the number of nights matches that which would be required to reach their destination. The birds will even flutter towards the end of their cage nearest their direction of travel. For garden warblers, which winter far down into Africa, this direction is southwest at first and then south after they have crossed the Sahara so that they avoid flying out into the Atlantic. Once again this change is found also in the caged birds (Figure 2.10): each autumn their fluttering takes on a less westerly component towards the end of the series of nights on which they are restless. This is true even of hand-raised birds which have not themselves ever been outside Europe.

The egg-rolling of a goose must be oriented in relation to the egg, and the migration of a warbler must carry it in the correct direction over the surface of the earth. But, beyond their role in orientation, external stimuli play little part in the pattern of these movements. The next chapter will consider more complex external stimuli and the role they play in behaviour, but here it is worth staying with orientational mechanisms for a little longer as these fall into various different categories which are quite helpful in thinking about movement patterns. Indeed, an early theory of behaviour was based upon classifying it by the various different forms of orientation involved.

The simplest of these mechanisms are called kineses, and these do not even involve the animal sensing the direction of the stimulus. If one side of a dish of water with some flatworms in it is covered with a dark cloth, the animals will be found to accumulate in the darkened area (Figure 2.11). But they do not do this because they are escaping from the light: it is because they move more slowly and turn more often in the dark. This is a kinesis because the direction of the light makes no difference. On the other hand, movements such as that of the fly larva which walks away from a source of light are referred to as taxes. These take various forms, the simplest being those directly towards or away from a source of stimulation, whatever this may be. Positive phototaxis takes an animal towards light; negative geotaxis away from the earth's gravitational pull.

One stage more complicated than this is a movement that is at an angle to the stimulus rather than straight towards or away from it. Homing pigeons sometimes use the sun to tell them the compass direction of home, they then fly at the angle to it appropriate to reaching their destination. This is not just a matter of taking up an angle, however, for the sun moves through the sky and they must compensate for this movement if they are to keep a constant direction. For example, if they are in the northern hemisphere and home is south, the sun should be to their left early in the morning, ahead at noon and over to the right at dusk. In other words, they must know what time it is as well as the direction in which they wish to travel: a not inconsiderable feat. In many ways it is rather easier for the warblers which migrate down to Africa, for they fly almost entirely at night. One outside cue that they use to guide them is the pattern of stars in the sky, and especially the small cluster of stars round Polaris, the pole star. These are useful because, to a bird looking

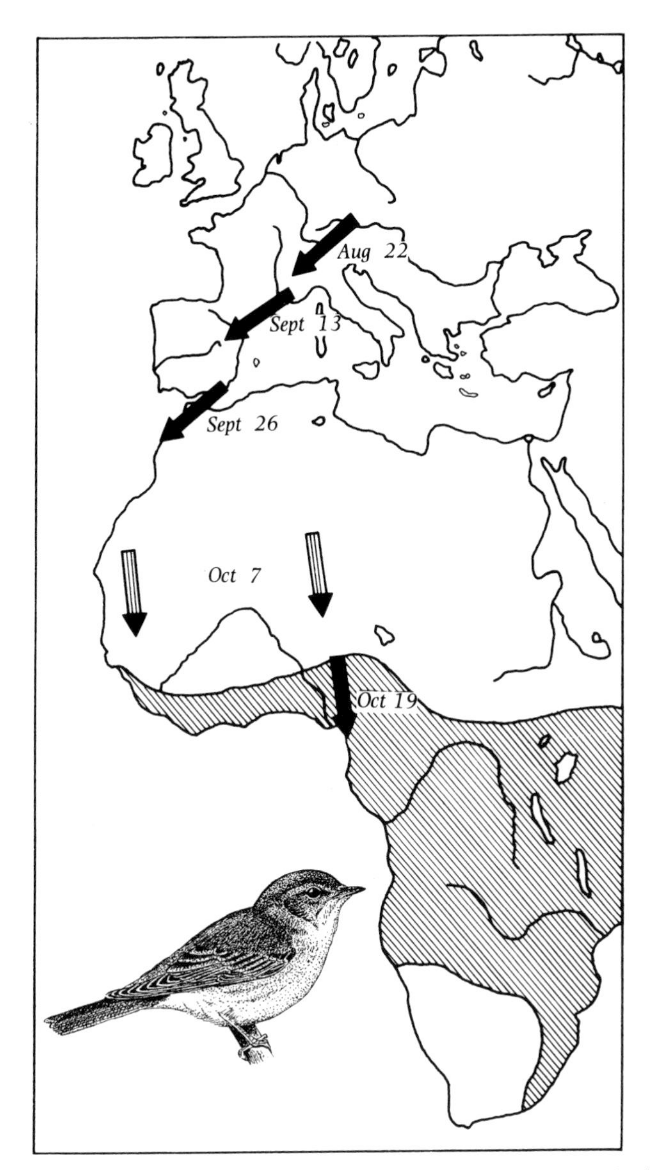

Figure 2.10. The migratory pattern of garden warblers which nest in Europe and then move to southern Africa in winter. The arrows on the map show the preferred directions of hand reared birds kept in the laboratory and tested on the dates when free-living ones would be at the locations shown. As autumn progresses their preference moves from south-west to south in keeping with the route they follow in the wild. (After Gwinner, E. & Wiltschko, W., 1978. *J. comp. Physiol.* **125**, 267–273).

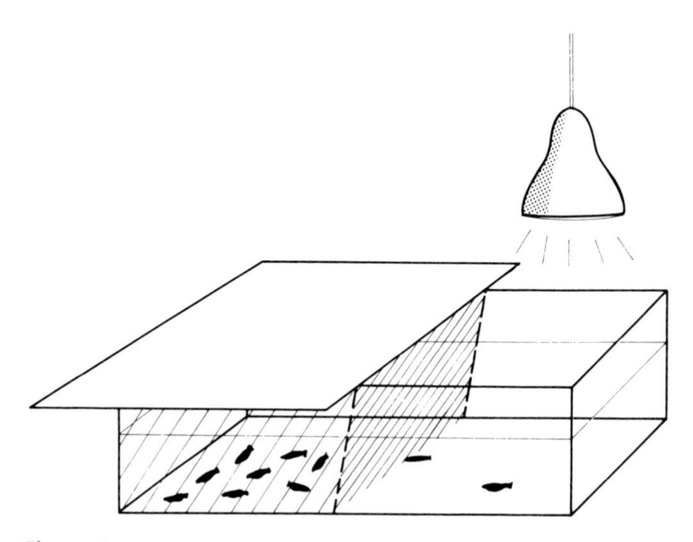

Figure 2.11. Flatworms in a half-covered tank gather in the darkened area not because they move away from light but because they move more slowly and turn more often in the dark than in the light.

upwards, the star pattern of the sky rotates round them and they remain in the same position regardless of the season or the time of night. A bird wishing to fly south need only fly away from them and does not require to know what time it is. Small birds also use the earth's magnetic field, in ways that are not yet fully understood, especially as far as the sensory mechanisms involved are concerned. An interesting example here is the European robin, which uses the angle of dip of the field to detect which direction is north and which south (Figure 2.12). A compass needle held vertically points horizontal to the ground at the equator and straight down at the north pole. As robins do not cross the equator they can be sure, without even knowing the direction of the field, where north is because the lines of force point downwards towards it. Perhaps it may seem curious that they do not have to know the polarity of the field, in other words which end of a compass needle is which. However, a good reason for this is that which of the earth's poles is which has switched every so often during geological time: though compass needles point north today, there have been periods when they would have pointed south. Thus, if robins depended on polarity, and a switch occurred they might all fly off to the north pole in autumn!

A final form of orientational mechanism is the way in which animals use landmarks to find their way around. This involves sensing the complicated patterns and relationships of objects in the world around them rather than just

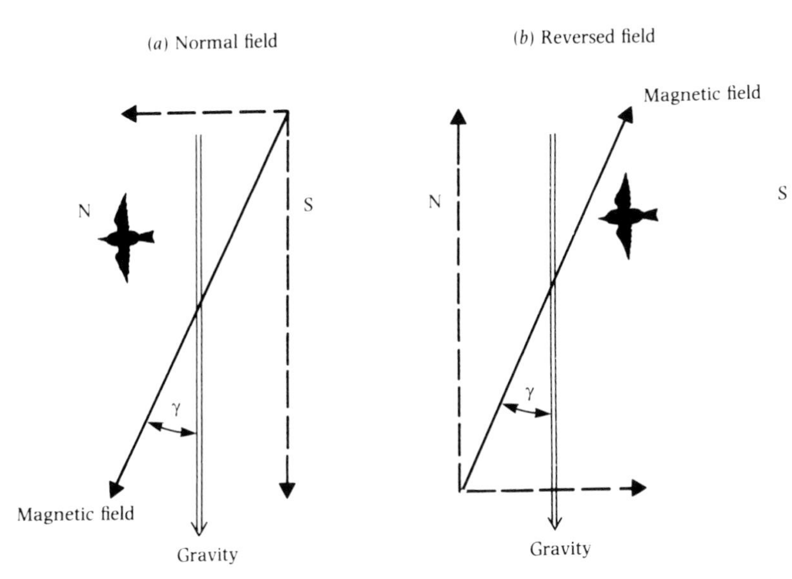

Figure 2.12. European robins move north in the spring and the direction they take is influenced by the earth's magnetic field. In the northern hemisphere this points down, the angle of dip depending on the latitude, and the birds find north by taking the smallest angle between the field and gravity (gamma in *a*). That it is this rather than the polarity of the field that they use has been shown by experiments in which the direction is reversed (*b*). Here the birds still orientate by the smallest angle but now go the opposite way from the direction of the field. (After Wiltschko, W. & Wiltschko, R., 1972. *Science* **176**, 62–4.)

the direction of this or that stimulus. It thus involves sophisticated sensory systems and brain mechanisms such as will be discussed in the next chapter. The navigation of robins or of homing pigeons may get them close to their destination but could hardly land them in the same bush as they nested last year or on the roof of their loft from hundreds of miles away. Knowing the features of their home environment undoubtedly helps them to reach the exact place after they have arrived in the rough area using their skills at long distance navigation. Indeed orientation by landmarks is an important capacity in most animals whether or not their movements normally take them far from home. They must move around their range to find food and water, shelter and mates. It is essential that an animal has a detailed knowledge of the places where each of these may be found and how they may be reached from wherever it happens to be. For example, imagine a rabbit feeding in the open when a hawk flies over. Its chances of survival depend crucially on its ability to reach the safety of a burrow within seconds. In some sense it must therefore have a map of its environment in its head and a knowledge of just where it is on that map: not surprisingly it is an important function of the nervous system to build up such maps.

2.5 Conclusion

The behaviour patterns shown by animals tend to be typical of their species and, while stereotyped, especially in the case of courtship displays, they are much more variable than originally thought. In general, while the simplest reflexes do not even involve the brain, more complex actions tend to be patterned centrally with feedback from the periphery providing modulation. Feedback is indeed an important aspect of most animal movement: each stage of an action is usually based on the outcome of the last, with sensory input playing a crucial part, especially in the case of orientation. The next chapter will take us on to consider sensory input further: just what sort of view of the world do their senses give to animals?

3

Sensory systems

Whether it is during the long distance migrations discussed at the end of the last chapter, or when trying to decide if the sounds made by a courting male belong to the right species, or in tasting food to see if it is palatable or poisonous, animals must keep constant check on their environment through the medium of their senses. These provide them with windows out onto the world in which they live, but they are windows whose features vary greatly from sense to sense and from species to species.

In humans, the two senses of vision and hearing are particularly finely developed, and we tend to think of these, rather than touch, smell or taste, as our primary contact with the outside. It is certainly marvellous the detail with which they provide us, especially when one considers that the scene one is looking at and the sounds one is listening to are not there inside one's head at all, but have been reconstituted from a stream of nerve impulses coming from the eyes or ears. These nerve spikes do not vary in size and shape to carry information, but are simple all or nothing signals. Which nerve fibre from the eye they pass along indicates the position, and sometimes also the colour, of the object to which they are responding; the rate at which they are sent indicates how bright it is. Similarly, nerve cells from the ear are stimulated only by particular tones, and the rate of firing they show indicates how loud these are. With literally thousands of receptor cells, each tuned in a different manner, and with two ears to help us to detect whereabouts different sounds come from, our brains are presented with all the information to reconstitute the sound, be it Bach or the Beatles. It is then up to our psychology whether we appreciate it or not!

Many other species, especially birds, as their bright colours and varied songs

Figure 3.1. A barn owl swoops down to capture a mouse, its movement caught in a series of flashes. The hearing of these birds is extremely acute and they can localise a source of sound in three dimensions so accurately that they have no difficulty in catching prey in pitch dark (photograph by M. Konishi).

suggest, also rely mainly on sight and sound, but their senses often differ from our ones in acuity and in frequency range. For example, barn owls have remarkably acute hearing which enables them to locate sounds much precisely than we can do (Figure 3.1). Like ourselves most birds can hear up to around 20,000 cycles per second, the hearing of some animals, of which bats are the most famous, goes up far beyond this to around 120,000 cps. Bats use this ability mainly for echolocation, which involves hearing the high pitched sounds they themselves produce when they bounce back off objects and so working out where the objects are. Just as we are unable to hear the 'ultrasounds' that bats use for this purpose, we are also unable to see some of the colours that other animals can sense, those with wavelengths shorter than violet or longer than deep red. Insects are sensitive to ultraviolet light and can, as a result, be attracted to flowers by colours and patterns on them that we simply cannot see (Figure 3.2). Many birds are also now known to see in the ultraviolet: compared with ourselves they have an extra type of colour receptor (cone) which is sensitive in that range. While it is striking enough to us, the iridescent plumage of starlings has additional components in the ultraviolet,

Figure 3.2. To our eyes the patterns on some flowers are not especially striking as in the marsh marigold shown on the left, but when photographed under ultraviolet light (right) the pattern is stronger and new features emerge. This is because the function of the patterns is to signal to insects and their vision stretches into the ultraviolet. (After Eisner, T. *et al.* 1969. *Science* **166**, 1172–4.)

and female starlings have been found to prefer males from which these have not been filtered out.

Despite these cases where sensitivity lies outside our range, sights and sounds are familiar to us, and we can get a reasonable impression of their impact on animals. It is more difficult for us to have any idea of the uses to which animals put the senses which we use less. The complex world of smells at the foot of a lamp-post may well be, to a dog, as exciting and full of information as is a magnificent painting or piece of music to us. Our reliance on touch is so slight that it is also hard for us to gain a 'feel' for the experience of an octopus as it wraps all its tentacles around an object to explore its shape. Even more difficult are the cases where animals possess senses we do not have or at least have no awareness of possessing. The rattlesnake uses two pit organs to sense the heat coming from its prey and, by measuring the differences between them as we do those between our ears when we localise sounds, can build up a map of the heat around it to locate the prey exactly. Some fish have electric organs with which they set up a field around them like the lines of force around a magnet (Figure 3.3). They use changes that occur in this field to locate objects and they themselves can alter the field as a way of signalling to one another. Lastly, there is the magnetic sense referred to in the last chapter, and possessed by bacteria and bees as well as birds. In birds and bees this sense probably relies on organs containing small amounts of magnetic material which are affected by the earth's magnetic field and so can be used by the animal to help it find its position.

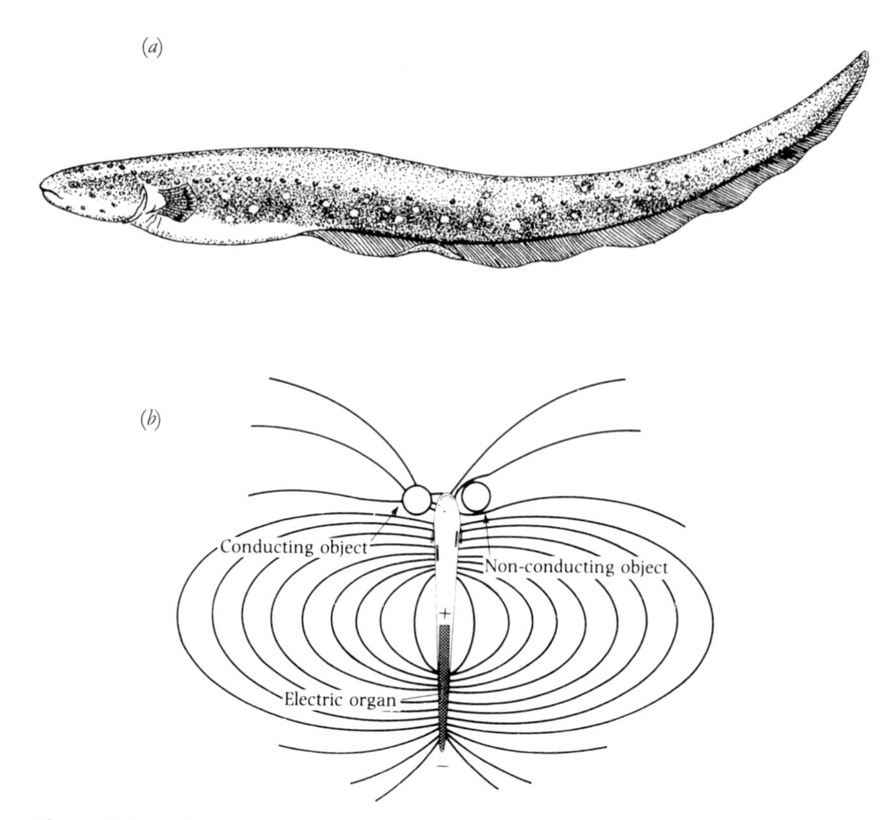

Figure 3.3. (*a*) The electric fish *Electrophorus electricticus*, showing the position of its electric organ in strips spreading forward from the tail. The receptors are largely concentrated in the head (*b*) The fields of force around an electric fish (in this case *Gymnarchus niloticus*) as altered by the presence of an object of high conductivity (left) and by one of low conductivity (right). (After Lissmann, H.W. & Machin, K.E., 1958. *J. exp. Biol.* **35**, 451–86.)

The diversity of sensory systems is a fascinating subject in its own right, but our task here is to extract some general principles about how the information they provide to the nervous systems of animals affects behaviour. We will take most of our examples from the senses which have been most studied because they are the ones we ourselves use the most: vision and hearing.

3.1 Sign stimuli and releasers

When a female stickleback in breeding condition enters the territory of a male that has completed his nest he shows the zig-zag dance, but when another

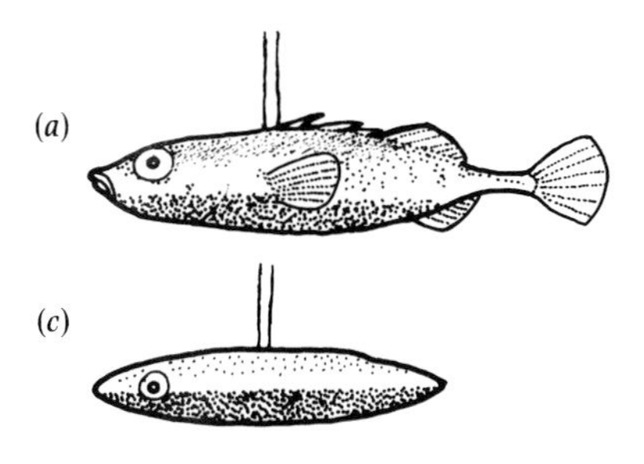

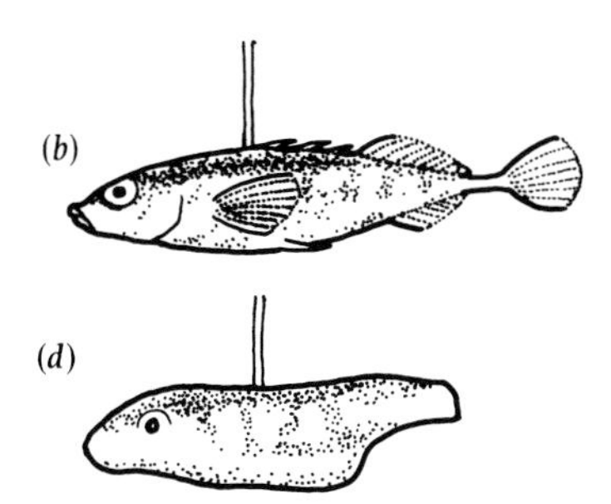

Figure 3.4. Three-spined stickleback models as used by Niko Tinbergen. That of a normal male (*a*) and an unripe female (*b*) are very similar, except for the red colour on the underside of the male. The details are not necessary for males to respond appropriately, however, as they will attack a crude model provided this has a red underside (*c*) and court one with a swelling beneath (*d*). (After Tinbergen, N. 1951. *The Study of Instinct.* Clarendon Press, Oxford.)

male appears he behaves quite differently, showing the head-down threat posture. How does he tell the difference? In a classic series of experiments Niko Tinbergen used various models, including those shown in Figure 3.4, to explore this question and found that the most important feature stimulating the zig-zag dance was the swollen belly of a ripe female, and that the one leading to threat was the red on the underside of the rival male. Sausage shaped models lacking fins, gills, spines, tail and eyes were responded to with courtship provided that they had a bump beneath like a female ready to spawn. As far as threat was concerned, Tinbergen even described how the males in his aquarium tanks would sometimes show the head-down posture when a red post office van drove past the window! Clearly, the colour red on its own was enough to give the display.

Experiments like these show that some features of a stimulus may be sufficient to bring about a particular pattern of behaviour. The general term for such features is 'sign stimuli', indicating that the animal attends particularly to them rather than to other details which may, on the face of it, be equally striking. In the stickleback case we can see why this might be so: the critical difference between a male and a female intruder is that the male has a red underside; that between a ripe and an unripe female lies in whether her belly is swollen. It seems appropriate therefore that natural selection has made the territory owner especially responsive to these points. In some cases there is reason to believe that selection has also changed the feature concerned to make it more effective as a signal. Eggs make female sticklebacks swollen anyway, and this is probably not the case with her. But the red of the male is clearly there because it acts as a signal to other males to keep out and to females to approach.

Sign stimuli such as these, which have arisen during evolution specially to function as signals from one animal to another, are known as releasers. Examples are numerous, from the colour patterns of butterflies and coral reef fishes, to the songs of birds and of cicadas, and to the smells with which mammals mark their territories and moths attract their mates. Releasers act in communication, and so we will return to them again in Chapter 8. They are, in effect, a sub-set of sign stimuli with this particular rôle. But many of the other sign stimuli to which animals respond are not releasers in this strict sense but simply aspects of the outside world which provide valuable cues to them and so alter their behaviour. The pattern of stars in the sky or the sound of a mouse rustling in the grass may influence the behaviour of a hawk, but they are certainly not signals sent to it specially to do so. Strictly speaking these are sign stimuli but not releasers, though the two terms are often used interchangeably partly because it is not easy to tell with some features of an animal whether they are specially adapted to act as signals. Does the belly of the female fish appear to bulge more than the eggs would naturally make it do?

As with fixed action patterns, the early ethologists had some very clear ideas about the nature of sign stimuli, many of which have had to be modified as a result of later work. One evocative analogy was with a lock and key mechanism. The sign stimulus was seen as like a key which turned a lock somewhere inside the animal, referred to rather grandly as the 'Innate Releasing Mechanism' (IRM), and so caused the animal to perform the appropriate action. Whether a single mechanism is usually involved in producing such responses, whether it releases the behaviour more or less automatically, and whether it is in any sense inborn, have all been questioned in particular cases by later work, so the expression IRM has fallen from use. But there is, nevertheless, no doubt that some of the ideas it encapsulates are useful. For example, animals certainly do show selective responsiveness, paying more attention to some aspects of a stimulus than to others. But in most cases the response is not simply a lock and key-like one, some features being all-important, the rest totally irrelevant. An example will help to illustrate this point.

3.2 Egg recognition in herring gulls

The egg of a herring gull is shaped like that of a hen but is rather larger. It tends to be green or brown in colour and to have a number of large dark blotches on it, especially at the blunt end. Like the goose referred to earlier,

Figure 3.5. A herring gull returns to its nest to find a choice between two egg models on the rim (From Baerends, G.P. & Drent, R.H. (eds.), 1982 *Behaviour* **82**, 1–416.)

the gull will retrieve one of its eggs if it rolls out of the nest, and the Dutch ethologist Gerard Baerends has used this behaviour in an ingenious and lengthy series of experiments to determine just what stimuli tell the gull that the object outside its nest is indeed an egg.

What Baerends did was to place two model eggs side by side outside the nest of a gull and watch from a hide to see which of them the gull rolled back first (see Figure 3.5). Over a series of thousands of such tests he varied the characteristics of the models so that he could see what were the preferences of the gulls. To avoid confusion, he changed only one feature at a time. For example, he tested the gulls with eggs varying in size from that of a pigeon's egg to that of an ostrich's, but kept their shape, their colour and their blotchiness the same throughout these tests. When examining the influence of shape he used models in the form of prisms, cylinders and rectangular blocks, as well as ones like eggs, but all of these were kept the same size and painted similarly so that he could see which shape was most effective.

One reason why it was necessary to carry out an enormous number of tests to decide which features were most important to the gulls, is that the birds showed a strong position preference, so that with two similar eggs some birds would almost invariably choose the left hand one and some the right. To allow for this, each bird had to be presented with many different pairs and their position had to be changed around in a systematic way. When such allowances were made, some very clear results emerged. For example, the birds tended to prefer the larger of two eggs, even if it was much bigger than their own and the other was the normal size. They also preferred egg models with many small speckles to the more natural ones with a few large blotches on them. Both these features probably arise because eggs which are larger and with more spots on them are more striking. Indeed it could simply be that they stimulate the eyes of the bird more and so are more likely to attract its attention.

In the case of other features the preference tended to be for those more like normal gull eggs. The birds preferred models shaped naturally to those in the form of cylinders or rectangular blocks. Amongst prism and block shaped models they chose those with rounded edges rather than ones with sharp corners. They preferred models with blotches darker than the background to those with lighter ones. They chose green, yellow or brown eggs rather than blue, red or grey ones. A slight difference between their colour preference and features of the normal egg is that they retrieved yellow egg models rather than brown ones, though the normal egg is green or brown but never yellow.

All these tests were carried out on models, and these obviously cannot be made to look exactly like the real egg. For instance, the wooden ones Baerends used lacked the textured surface typical of egg-shell. Nevertheless, a model could be made that the gulls would roll back in preference to one of their own eggs. It was 50% larger than normal, green, and with many tiny black speckles on it. This is a good example of what ethologists refer to as a 'supernormal' stimulus as it is more effective than the stimulus found in nature.

These results show that many factors contribute to the egg-rolling response. Though an egg may be retrieved even if its size, shape and colour are wildly different from the gull's own egg, all these features do have a rôle in making the animal more likely to respond. One feature may be enough, but many others do contribute. They add together, in what has sometimes been rather grandly called 'the law of heterogeneous summation'. At first sight this may seem rather different from the stickleback example, but it is not really. Although male sticklebacks may threaten any red object, without very extensive experiments one cannot determine whether red is all that matters or whether they would threaten more if the object was the right shape, had fins and a tail, spines and an iridescent blue eye.

What both sets of experiments do illustrate is that animals may respond differently to objects that look quite similar to each other and that certain

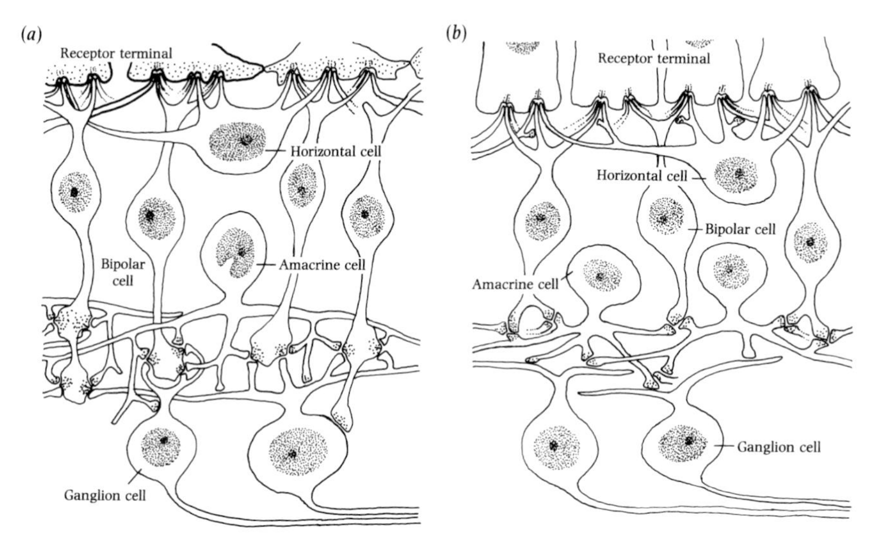

Figure 3.6. The cell organisation in the retina of a primate (*a*) and that of a frog (*b*). Similar classes of cells exist in both cases but, in the primate bipolar cells connect the receptors directly to the ganglion cells, which send axons to the brain, while the arrangement is more complex in frogs, which have no such direct connections. (Redrawn from Michael, C.R. 1969. *Scient. Amer.* **220**, 104–14.)

features of a stimulus may be more important than others in leading to their response. We must now consider the sensory basis of these differences and, in particular, the extent to which they have their origin in the responsiveness of the sense organs rather than in the brain. The sensory world of frogs and toads has been studied extensively, and this provides some excellent examples.

3.3 What the frog's eye tells the frog's brain

The title of this section is taken from a classic paper by Lettvin and his co-workers on the vision of amphibians. It might seem obvious that the animal's eye should tell its brain as much as possible about what the world around it looks like, but this is not what it does at all. In frogs and toads the eye is actually more complicated than it is in ourselves, with several layers of cells between the light receptors and the nerve fibres which travel from the eye to the brain (Figure 3.6). The ganglion cells are the last staging post before these fibres leave, and the diverse connections of the cell layers which precede them mean that each ganglion cell receives input from a large number of receptors.

Recording from ganglion cells shows that some of them respond best to very particular objects. For example, one class fires when a small rounded object passes into the field of view and will continue to do so if the object becomes stationary. Interestingly, this cell will not fire if such an object suddenly appears in front of the frog without being seen to move in; nor will the cell respond to a larger dark bar moved across in front of the eye. Because of their particular characteristics these cells in the retina of frogs have been aptly named 'bug detectors', and that is probably their exact function. Objects that cause them to fire are much more likely than other things to attract the attention of the frog as a result, and they are also much more likely to be food than are vertical rods or static spots. Broad edges moving into the frog's field of view, and sudden shadows moving across it, stimulate other types of ganglion cell, and these are probably concerned with recognition of much larger objects such as hawks and other predators. Thus the frog's eye does not tell the frog's brain about things that are irrelevant to it, but selects only the important features to which the animal should respond, such as stimuli likely to be edible and those likely to be dangerous. To the former the frog will flick out its tongue, which is very long and has a sticky end on which flies become impaled. To the latter the frog will show escape.

Despite the fact that the eye of a frog or toad clearly passes on only the most relevant information to its brain, so that only a very rudimentary picture of the world can be recreated there, not all the animal's reactions can be put down to which types of ganglion cell in its retina have been stimulated. For example, when a toad snaps at prey it must assess the distance away that this is so that it can extend its tongue by an appropriate amount. To do this requires binocular vision involving integration of information from the two eyes, and this is carried out in a part of the brain called the optic tectum which consists of two nuclei, one on either side (see Figure 3.7). The nerve fibres from each eye travel back to the optic tectum on the other side of the brain and the connections they make there are mapped out over the surface of the tectum in the same arrangement as are the positions in the eye from which they came. But each cell in the tectum is also connected across to one seeing the same point in space on the other side, as shown in Figure 3.7. These binocular cells can be used by the animal to assess distances.

The rôle of the brain in the selective responsiveness of amphibians has been shown particularly well by work on prey capture in toads by Jorg-Peter Ewert. The best stimulus for snapping at prey by a toad is a long thin dark bar, moving across its retina like a worm. If such a bar has a side projection added to it, so that it becomes L-shaped, the snapping response of the toad tends to be inhibited: the worm has become an 'anti-worm'. Ganglion cells in the retina

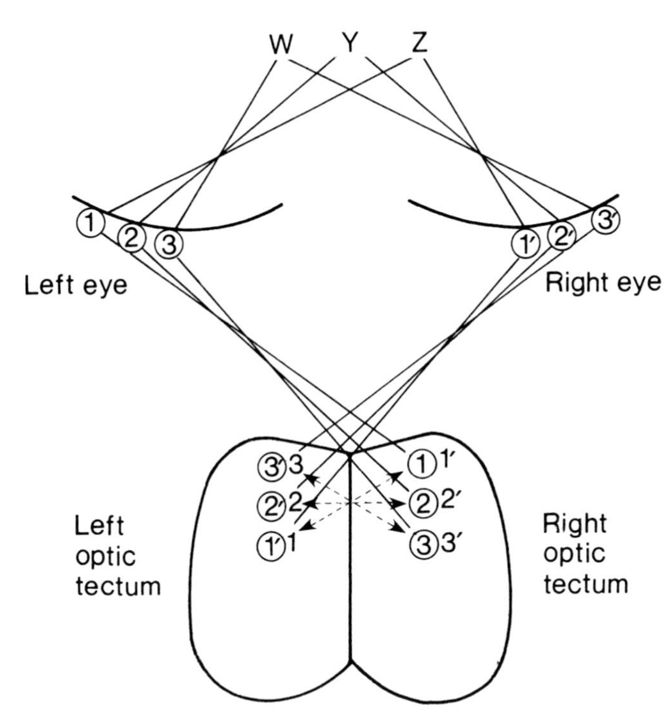

Figure 3.7. Binocular vision of the toad *Xenopus* enables it to assess distance and so capture prey effectively. Fibres from each eye pass to the optic tectum on the far side of the brain, but connections between the optic tecta link cells which receive the same visual input. Thus an object at Z in space stimulates point 1 on the retina of the left eye and point 1′ on the right. A cell in the left tectum which receives direct input from point 1′ also has an indirect connection via the other tectum to point 1, so laying the foundation for binocular vision. (After Keating, M., 1974. *Br. med. Bull.* **30**, 145–51.)

respond to objects of the right size and speed, but do not discriminate between worms and anti-worms. Instead, this discrimination takes place in the optic tectum: there are cells here which do distinguish between different types of stimuli, though no one of them has all the properties of prey recognition. It seems that the system depends on a network of nerve cells which receives input from the two eyes and from which output goes to the tongue musculature and so gives the response.

Turning now to the hearing of amphibians we find that the ears of frogs, like their eyes, are tuned to sounds of particular interest to them, in this case the calls of their own species. Figure 3.8 shows the detector mechanism which bullfrogs are thought to use. Some receptors in the ear are stimulated

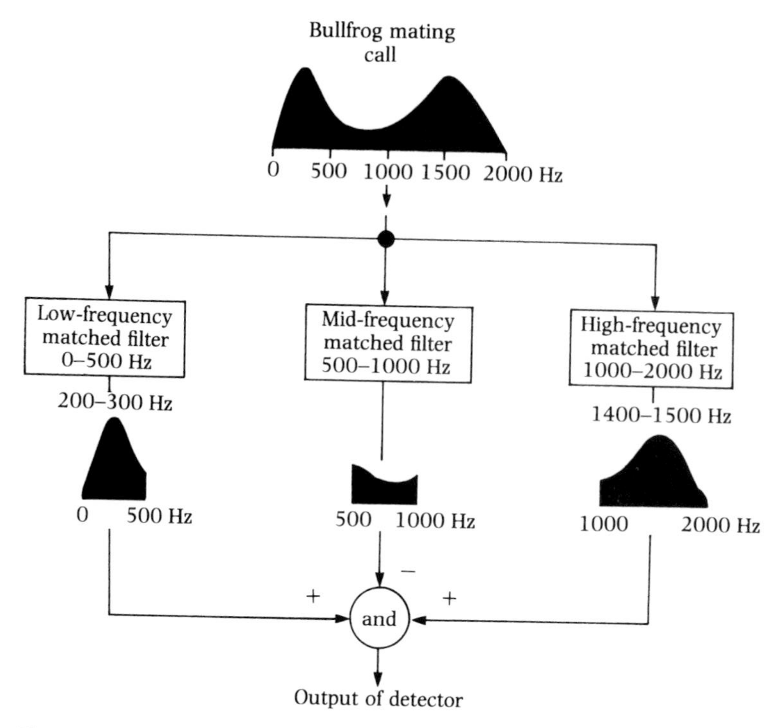

Figure 3.8. Diagrammatic representation of how bullfrogs detect each other's calls. The call has energy at two different frequencies and cells in the ear respond preferentially to sounds at two different peaks. Intermediate sounds, between 500 and 1000 Hz (cycles per second), suppress response. It is postulated that detection is based on three filters matched to the three frequency bands, high and low frequencies stimulating the detector while medium frequencies lower its output. The detector is an 'and' one: there must be energy at both peaks and there must not be energy in the trough before it will respond. (From Capranica, R.R. & Rose, G., 1983. In *Ethology & Behavioral Physiology*, ed. F. Huber & H. Markl, pp. 136–52. Springer-Verlag, Berlin.)

especially by tones of just over 1000 cycles per second, while others respond mainly to sounds at around 300 cps and their firing is suppressed by frequencies rather higher than this. So, a sound at the lower frequency makes the cell respond but adding another slightly higher pitched sound to the first will decrease its firing. This system matches the animal's hearing perfectly to the two energy peaks in the call of the male bullfrog.

This is a good example of a peripheral filter at work and shows how subtly the sounds animals produce and the senses of their hearers may be matched to each other. Indeed the match may be yet more precise. Cricket frogs produce higher pitched sounds and their calls show dialects, varying in frequency from place to place. But the frequency a male produces in a particular area is matched to that of the female's ear in the same place: 3750 cps in East Kansas, 3550 in New Jersey, 2900 in South Dakota. Another fine piece of matching between the calling of the male and the response of the female stems from the fact that amphibians are cold blooded. This means that on a cold day the male grey tree frog in Missouri produces fewer pulses per second than when it is warm which could obviously lead to problems in mate attraction. However, it transpires from experiments by Carl Gerhardt that the preference of the female varies in an exactly parallel fashion. The only problem comes when she is sitting in a place where the temperature is very different from where he is calling, for then she ignores him: but that particular circumstance only really arises when a prying scientist has changed the temperature of one of them!

3.4 Central and peripheral filters

From these examples it is clear that animals are endowed with the equivalent of filters, which let some stimuli past and so lead to behaviour appropriate to them, while excluding other stimuli which are ignored or treated differently. Sticklebacks do not attempt to court their rivals, nor do gulls try to roll their chicks into the nest when they run out of it. In some cases the filters appear to be properties of the sense organs while in others more central mechanisms are involved. We can ask somewhat similar questions here to those we considered about whether movement patterns are centrally or peripherally organised. Why is it, then, that some of the stimuli to which animals respond are simply those to which their sense organs are tuned, while others are much more complicated combinations of features, the recognition of which requires a great deal of processing by large numbers of nerve cells within the brain itself?

Some filtering obviously goes on in all sense organs as they do not respond to every conceivable stimulus. The fact that we cannot see the ultraviolet patterns on flowers is an example of this. But no one would doubt that, within their range of sensitivity, a great deal of raw information is passed by our eyes and ears back to the brain and that it is there that we decide what to do with it. The problem with more precise sensory filtering, like that involved in many of the examples from amphibia discussed above, is that it can only work where the sense concerned is used in responding to one or a few very specific stimuli. A good example of a sense organ tuned so that it can be used for two different things is in the ear of the cricket *Teleogryllus oceanicus*. Like other crickets, these animals show peaks of responding at frequencies matching those in their species song. But they also have units which are tuned to much higher frequencies and these probably enable the crickets to detect the ultrasonic cries of a bat which preys upon them. Their ears (which incidentally they keep in their elbows!) seem therefore well adapted for detecting both mates and predators.

Despite this example, it is hard to see just how a sense organ could be wired up to allow an animal to distinguish between many different sorts of stimulus. Precise peripheral filters are therefore most likely to be found in animals with few interests in life. The ears of frogs are probably largely concerned with hearing the rather stereotyped croaks of other members of their own species and have little rôle in detecting food or predators. Their eyes are largely concerned with the distinction between things to eat and things to run away from, though they certainly also have a rôle in mating once a female has been attracted to close range. However, the information they provide on mates is not very precise: male frogs and toads are notoriously unselective about the objects they will attempt to clasp during the mating season!

The uses to which herring gulls put their sense organs are much more varied and a great deal of the responsiveness that they show must, accordingly, be centrally determined. Although gulls do respond best to large eggs with many small spots on them, and this could easily be because such eggs stimulate their eyes most, their preference for other aspects is not so easily explained. The favourite colour of egg for the egg-rolling response is green, yet the feeding gull prefers objects which are brown and the gull chick, when pecking for food from its parent's beak, pecks the red spot on the beak and prefers red or blue to other colours. Preference for different colours under different circumstances is unlikely to be accounted for at the level of the sense organs, although many sense organs do have motor nerves going to them from the brain which are probably involved in adjusting sensitivity in various ways and could even swing different types of filter into action.

In birds and in mammals the retina is much less complicated than it is in lower vertebrates but the visual pathways in the brain are more extensive. Fibres from the retina pass to the optic tectum, known as the superior colliculus in mammals, and this seems to be involved in locating objects in space as it is in toads. However, there are also connections from the eyes up to the visual cortex of the brain, and here a great variety of different types of cells are found. Many of them respond to edges passing across the retina in particular orientations, and these cells occur in ordered columns where those near each other have similar preferences. Other cells are binocular, firing to stimuli at equivalent points on the two eyes. Yet other cells have been found with much more complicated and specific preferences: for example, some cortical cells in starlings fire only to a particular species-specific call, and some in monkeys will recognise a face.

While a male toad may confuse a wide variety of objects of about the right size and shape with potential mates, this is not usually a problem with gulls, rats or humans. Indeed these animals may be able to distinguish their mates from all other members of the species using subtle visual, auditory or olfactory cues. Gulls, rats and people are also omnivores and recognise all sorts of foods while rejecting many objects which look similar but are in fact inedible. The distinction between predators and similar species that are harmless can also be slight yet it creates few problems: many small mammals of the African plains will scurry away at the sight of an eagle yet feed unperturbed as a vulture passes over. In cases such as these, where many different stimuli have to be assessed and responded to appropriately, the recognition processes involved must depend on the computer power of the brain rather than on simple filters in the sense organs themselves.

3.5 Sensory exploitation

In Chapter 8 we will return in some detail to releasers and their effect on other animals, for this is the essence of communication. But here we will whet the appetite with the idea of sensory exploitation, introduced a few years ago by Michael Ryan. It is obvious that the display of an animal should come within the sensory range of those at whom it is directed. But, within that range, certain stimuli may be much more effective than others, for reasons quite unconnected to communication. If, for example, a species feeds on blue berries and its senses are tuned to this, perhaps males would be more striking and attractive to prospective mates if their releasers exploited this bias and incorporated the colour blue.

A number of examples of this process have come to light. In swordtails, females prefer males with larger swords, and this has been found to be true even in the closely related platys, which do not normally have swords, by artificially attaching one to males. A possible explanation here is that it may benefit females to mate with larger males, and some species have matched this preference more cheaply by growing a sword than they could do by getting bigger. In another example, Ryan and his colleagues described how females of the group to which the túngara frog belongs prefer deep sounds and, in the túngara frog itself, the male mating call includes a 'chuck' component

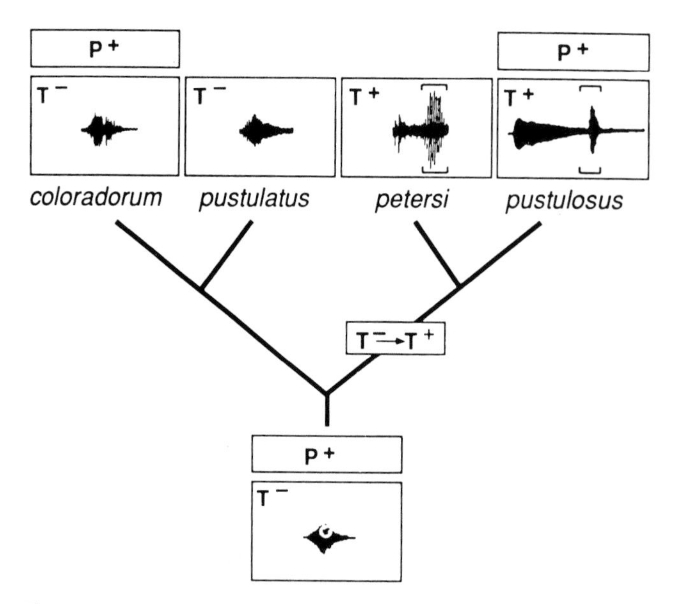

Figure 3.9. Male song and female preferences in four species of frogs of the genus *Physalaemus*. Only the closely related *P. petersi* and *P. pustulosus* (the túngara frog) add 'chucks' to their calls, as shown by the bracketed sections of the oscillograms, and so this probably evolved in their immediate ancestor. Females of both *P. coloradorum* and *P. pustulatus* prefer calls with chucks. This suggests that the preference was present in the common ancestor of all four species, before the chuck itself evolved. The oscillogram at the bottom is typical of the other calls in the genus. (T^-, no male chuck; T^+, male chuck; P^+, female preference for the chuck.) (From Kirkpatrick, M. & Ryan, M.J. 1991. *Nature* **350**, 33–8).

(Figure 3.9). The larger the male, the deeper this can be, and females prefer larger males. However, closely related species, while having the same female preference, have evolved no chuck. It appears that this evolution has taken place in the túngara frog to match what was a pre-existing preference in females of the group.

A final entertaining, but no doubt surprising, example, concerns the evolution of the teddy bear. Konrad Lorenz argued that adult animals have what he called a 'cute instinct', being particularly attracted to nurturing youngsters that have certain features in common, especially a large head, with high forehead, big eyes and bulging cheeks, as well as a small nose and short legs (Figure 3.10). Such features are found in the young of many species (and in a variety of Walt Disney characters!). On the other hand, the teddy bear did not start out that way, but originally had long legs and a rather pointed nose. Robert Hinde and L.A. Barden measured a lot of teddy bears and found that, as this century had

Figure 3.10. The young of many species have high rounded foreheads, small noses, and large eyes and ears. Lorenz suggested that these features were particularly attractive to adults, which therefore tended to nurture and protect them. (Redrawn from Lorenz, K.Z. 1943. *Z. Tierpsychol.* **5**, 235–409.)

progressed, their noses had become stubbier and their foreheads higher, so that they had evolved to fit in more and more with the features Lorenz described. Recent evidence suggests that small children do not appreciate these 'baby characteristics', so it is human adults that are being subjected to the sensory (and commercial) exploitation.

3.6 Conclusion

Animals have a great variety of senses, some of them quite alien to us, and even in those we share their range can often extend beyond our own. Where a sense is used only in one situation, like mate recognition, it may be very precisely tuned to the characteristics of the relevant stimulus, often within the sense organ itself. But, where senses have a wider range of uses, this cannot occur, and in this case central mechanisms must be involved in linking stimuli to the appropriate responses. Natural selection has led animals to be more sensitive to some stimuli than to others but, while such sign stimuli may be adequate to elicit a response, the case of the herring gull's egg shows that many features may sum up to enhance it. Releasers are examples of sign stimuli which act as signals to other individuals, a topic taken further in our discussion of Communication in Chapter 8.

4

Motivation

When it encounters a particular stimulus, an animal does not always show the same response. As we saw in Chapter 2, even the simplest of reflexes shows fatigue and must then be allowed a period of recovery before it can be elicited again. However, few of the changes in responsiveness shown by animals can be put down to fatigue. A male stickleback on his own in a tank and shown a model female for a brief period every day may sometimes court it with great vigour, yet on other occasions he ignores it. Anyone who studies animal behaviour soon comes to realise that it varies enormously from time to time, and this is frustrating if one hopes that animals will always behave in the same way. But, on the other hand, discovering just what it is that leads them to behave differently from one time to another presents some very interesting problems that are well worth studying in their own right. These are the problems of motivation or, in other words, the mechanisms leading animals to do what they do when they do it.

What motivates an animal from within is obviously a much more difficult subject to come to grips with than the stimuli from the outside world which affect its behaviour, and it can be studied in two very different ways. One is to look inside and see what is happening there, for example by stimulating nerves or recording from them. The hope here is to find the pathways between the sense organs and the muscles, and the physiological factors which affect them and so influence the behaviour. This approach has had some notable successes, especially with relatively simple systems, like that of silkmoth eclosion discussed in Chapter 2. But it would be a formidable task to understand more complex behaviour in these terms, and there is also the point that systems tend to have emergent properties so that the whole is more than just the sum of its parts.

The alternative approach, more conventional amongst those interested in behaviour, is to treat the animal as a 'black box'. This involves changing the inputs to it in various different ways, for example by depriving it of food or introducing a rival to it, and seeing how that alters behaviour, rather than looking inside to examine the underlying mechanisms. Just as one need not understand how a computer works to program it and to follow the output it provides, a lot can be found out about motivation without any knowledge of the hardware responsible inside the animal. Indeed, it can be argued that the 'black box' approach is an important preliminary in any case, to describe the phenomena that neurobiologists then need to examine. This chapter will illustrate some of the approaches that have been used to study motivation in this way.

In thinking about motivation it is worth considering two rather different aspects separately. One of these is how the tendency to behave in a particular way fluctuates over time, and this involves looking at different types of behaviour and the factors which affect them without much concern for other activities. As an example of this, one might study how eating is patterned in time and how it is affected by periods of food deprivation. The second approach is to examine how the animal decides which of the many behaviour patterns at its disposal it will perform. In this case one might look to see whether a rat that is hungry and also thirsty decides to eat or drink first. We will consider these two approaches in turn, starting with a simple model that was aimed at trying to account for changes in the tendency of animals to behave in one particular way.

4.1 A model of motivation

We all experience hunger at some times and not at others and so we realise that the urge to eat is not constant. The sight of a delicious meal is on some occasions mouth-watering and on others of no interest at all. Once again we owe to Konrad Lorenz a clear theory of how such changes might come about which helps one to think about the problem. His theory was put forward in the form of the model shown in Figure 4.1, the word 'model' in this sense meaning a simple scheme which is proposed to work in a similar way to the real system. Lorenz's theory is known either pretentiously as his 'psychohydraulic' model or, in a more down to earth way, as 'Lorenz's water closet'.

This model is not something one would look for inside an animal! It is what one calls an 'as if' model: the animal may behave as if it had such a system for organising its behaviour within it. The theory helps one to think about the

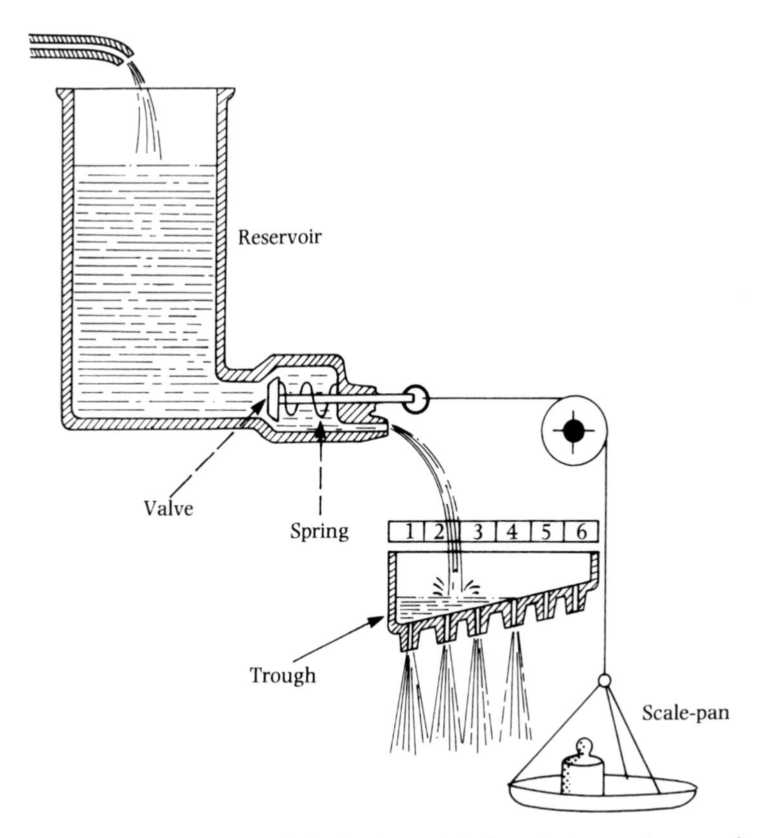

Figure 4.1. Lorenz's psychohydraulic model of motivation. Action specific energy is represented by water, which accumulates progressively in a reservoir when the behaviour concerned is not being expressed. The behaviour pattern occurs when the water passes out of the reservoir into the trough beneath. Higher threshold aspects of the behaviour (numbered 4, 5 and 6) are only shown when a lot of water is passing into the trough. The valve is so arranged that it is opened by the combined effect of the water in the reservoir (action-specific energy) and of weights on a scale pan, which represent the adequacy of external stimulation. (After Lorenz, K.Z., 1950. *Symp. soc. exp. Biol.* **4**, 221–68.)

problem and design experiments to see how the system really works rather than proposing specific mechanisms. Lorenz supposed that different actions depended for their appearance on a supply of 'action specific energy' which accumulated with time since the animal last behaved that way and was used up as it performed the act. He visualised this as water accumulating in a tank out of which it could only escape through a valve at the bottom. The valve was, however, a rather strange one. It could be opened either by the water pushing from within, or by a string attached to a scale-pan pulling from outside. To Lorenz, weights on this scale-pan were the equivalent of stimuli leading to the behaviour: the more appropriate the stimulation, the heavier the weights and the more likely the valve was to be opened.

Lorenz's model has some interesting properties which can be compared with real behaviour. First, the longer since the behaviour was last performed the more action specific energy will have accumulated and the more likely it is that the behaviour will appear. Second, the model suggests that the accumulation of action specific energy will lead the behaviour to occur even if the stimuli present are slight (the weights on the scale-pan are very light). Ultimately, Lorenz argued, behaviour of which an animal has long been deprived will appear as a 'vacuum activity' with no stimulus present at all. At the opposite extreme was a third feature, exhaustion: this occurred, according to Lorenz, when an activity had been stimulated so often that the animal had run out of this energy. Then, there is also a fourth idea incorporated in the model. The way the valve works suggests a particular relationship between internal factors and external stimuli: provided that there is some action specific energy, the push of this and the pull of external stimuli will *add up* to give rise to the behaviour, rather than being multiplied together or related in some more complicated way.

How does real behaviour match up to these four suggestions that Lorenz made? Let us consider each in turn.

4.1.1 Accumulation of energy

The nervous systems of animals have no stores of energy within them like Lorenz proposed, and this is obviously not a realistic aspect of the model. Nevertheless, animals might behave as if they did and, if so, we would expect them to become more likely to perform a particular action with the passage of time since they last did so. Do they do this?

There is no doubt that certain aspects of behaviour do become more likely the longer the gap since they last appeared (see Figure 4.2). Feeding and drinking are the most obvious of examples, as we all know from our personal experience. There are very good reasons for this: nutrients are used up by the body and water is lost constantly from it, so both must be replenished and the need to do so will rise with the interval since the last meal or the last drink. But a lot of the other activities which animals perform are not concerned with regulating aspects of physiology, so there is no reason to think in advance that they might become more likely with time, and they are quite often found not to do so.

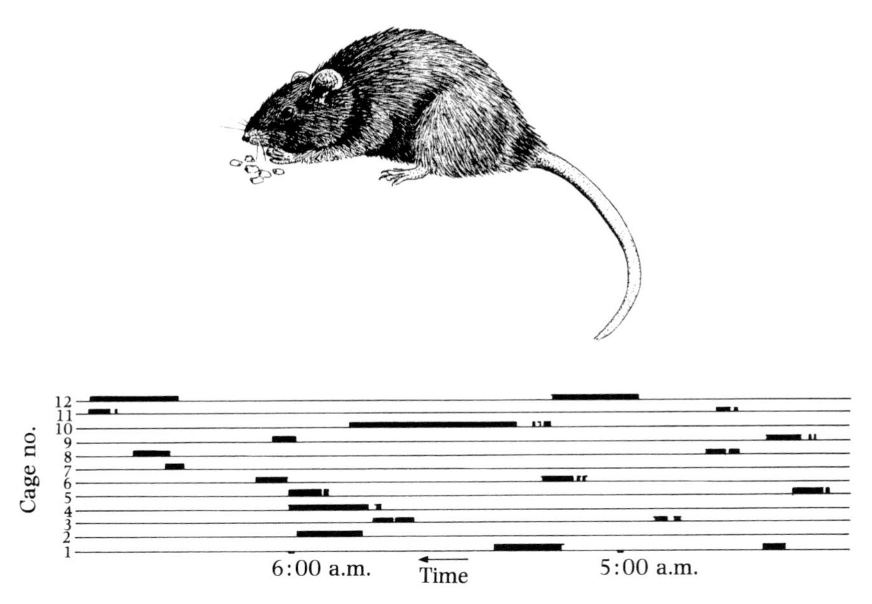

Figure 4.2. Traces from a pen recorder to show the feeding behaviour of 12 rats over a period of about 2 hours. The animals tend not to nibble at food the whole time, but pattern their feeding into meals, taking one of these at fairly regular intervals. Thus, the longer it is since its last meal, the more likely it is that a rat will eat. (After Thomas, D.W. & Mayer, J., 1968. *J. comp. physiol. Psychol.* **66**, 642–53.)

Many behaviour patterns, such as singing in birds, mating in fish or exploration in rodents, could be used to illustrate this point. However, a particularly appropriate one to take is aggression. Although he did not mention his psychohydraulic model, in his book *On Aggression*, Konrad Lorenz obviously had it in mind. He suggested that aggression is an innate drive which rises with time and must somehow be expended. His view of innate behaviour was that it is inevitable (a point taken up in the next chapter) and the only alternative is to sidetrack it into harmless channels: for human aggression he suggested that sport may play such a role, helping us to get rid of otherwise destructive urges.

Leaving aside the issue of whether sportsmen are less aggressive than other people as a result of their activities, there are a great many objections to these views. Some of them concern whether the urge to behave aggressively does accumulate in an inevitable fashion. One of the few cases where it seems to do so is where an animal is isolated, for example a mouse placed in a cage on its own. The longer it has been in isolation the more it will fight another mouse which is put in with it. But this could simply be that, when in company, it habituates to the other animals with it and it needs some days on its own to

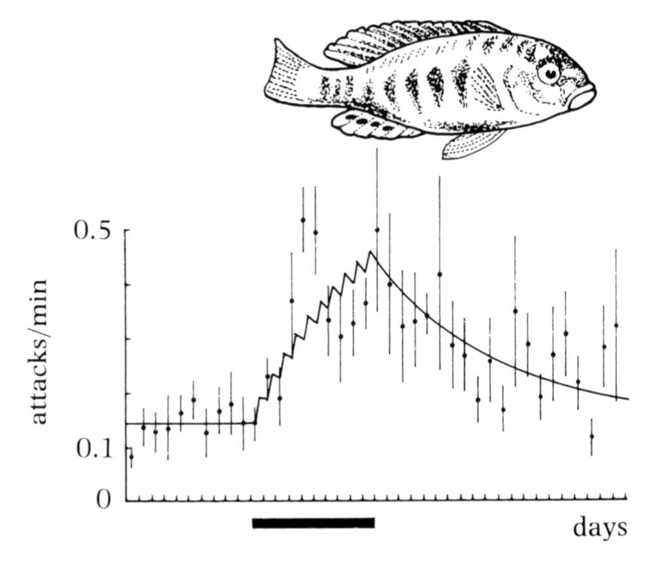

Figure 4.3. Males of the cichlid fish *Haplochromis burtoni* seldom attack small fish kept in their tank but, if a model of another large male is put in with them for 10 days (black bar), their attack rate slowly rises, then falls back again gradually after the model is removed. Thus they do not attack most as soon as the model is presented, as would be expected if they accumulated action specific energy in its absence, but instead need to be stimulated by a period of its presence before their attack rate rises. (From Heiligenberg, W. & Kramer, U. 1972. *J. comp. Physiol.* **77**, 332–40.)

recover its tendency to fight them. In fact some fish fight *less* after isolation and need stimulation from others over a period before the urge to fight returns (Figure 4.3): in this case contact with others probably leads to the gradual build-up of a hormone which makes aggression more likely.

Many factors affect aggressive behaviour, such as hormones and shortage of food on the inside, and presence of rivals and contested resources on the outside. Furthermore, aggression is itself a complex of different actions which may be quite differently caused, though they are superficially similar. It is not an easy matter to decide the extent to which a hawk killing a mouse, a cornered subordinate rat defending itself against a dominant, two fighting cocks locked in a struggle, and a mother duck defending her chicks from the unwelcome attentions of a gull, are actually motivated in a similar way. Aggression serves many different functions within one species (Figure 4.4), and its uses and the situations in which it appears vary between species, so that the mechanisms underlying it are also likely to vary. Evolution matches behaviour marvellously well to an animal's particular environment and way of life.

Figure 4.4. A cat cornered by a dog shows defensive behaviour, fighting back as best it can (*a*). Cats will also attack mice they encounter, here pouncing and showing other predatory actions as well as using their teeth and claws (*b*). In territorial disputes cats may fight intruders of their own species (*c*). Many of the actions involved in these different situations are similar. Although all of them involve attempting to inflict harm on other individuals, they probably have very different causes, so it is not useful to think of them all simply as examples of 'aggression'.

So it should be no surprise that the organisation of aggression varies a lot between species just as feeding differs between the lion that kills and eats every few days and the wildebeest that must crop its plant food for most of its waking hours. Likewise, there is no reason why the urge to behave aggressively should mount with time like hunger or thirst. Indeed, it is rather surprising that Konrad Lorenz should have thought this, given that he was a biologist with great respect for the power of evolution to adapt behaviour to an animal's exact requirements.

A final aspect of the accumulating energy idea concerns the extent to which an animal needs to perform the action as opposed to needing to achieve its results. Must a hungry animal make the appropriate number of movements, as Lorenz's model would propose, or does it just need to have the right amount of food inside it? Does an aggressive animal have to perform a certain amount of fighting, or will it cease to be aggressive when its rival is repelled? In Chapter 2 we saw that animals tend to regulate their behaviour as they go along in the light of its consequences: a process called feedback. The dog does not rush at the place where the rabbit used to be but follows an arc so as to close upon it, changing its course as required. Unless its muscles are fatigued, it does not stop running before it reaches the rabbit, nor does it carry on after it has caught it. This is quite different from the action specific energy idea, which proposed that the animal should perform the amount of behaviour for which it has accumulated energy regardless of its consequences. This would suggest that an eating animal should chew and swallow a set number of times even if an experimenter had surreptitiously raised the glucose in its blood or filled its stomach with food. In fact, animals do not generally do that: unlike the egg-rolling goose, they respond to feedback from their actions. They drink until receptors tell them that they have taken in enough water and they eat until they have had enough food, rather than showing a fixed amount of behaviour depending on the time since they last ate or drank.

4.1.2 Vacuum activities

There are few examples of the idea that animals deprived of an opportunity to perform an action might eventually show it even in the absence of all stimuli. Lorenz himself described a pet starling that would flutter up to the ceiling to catch a non-existent fly, but it is hard to be certain that there was nothing there to which the bird was reacting. There is no doubt that behaviour can sometimes show stimulus generalisation so that animals deprived of the usual stimulus will show the behaviour to a much less adequate one. In zoos, they will often mate with the wrong species when their own is not available: lions and tigers, for example, will mate to produce 'ligers'. As we all know, the hungrier one is the less tasty the food one is prepared to eat (indeed, if very hungry, I suspect even I might eat rice pudding!). Thus, generalising to less adequate stimuli is a reality, but there is not so much evidence for behaviour being shown in the total absence of any appropriate stimulus, as the

Figure 4.5. A European robin attacks the spot in mid-air where previously a stuffed rival had been placed.

vacuum activity idea suggests. Perhaps the best is an account by the ornithologist David Lack of the behaviour of a European robin he had just finished testing with a stuffed bird of its own species, a potent releaser of aggression for these birds. As he removed the specimen and walked off he chanced to look back and saw the robin fluttering around, singing and delivering pecks to the position in mid-air where its apparent rival had been (Figure 4.5). Here the appropriate stimulus had clearly been removed, but the animal was certainly not suffering from deprivation of the opportunity to behave aggressively! Indeed, the closest parallel may be with after-discharge, like that shown by Sherrington's dogs, which went on scratching after the stimulus was suddenly removed.

4.1.3 Exhaustion

In Chapter 2 we discussed how reflexes might cease to be performed through muscular fatigue, through sensory adaptation, or through some process of neural exhaustion which took time to recover. Could the last of these simply be like running out of action specific energy? Several detailed studies cast some doubt on such a simple view. A good one with which to start is a study of water boatmen by a Dutch ethologist, H. Wolda (Figure 4.6).

Water boatmen are insects which lie suspended upside down just beneath the surface of a pond waiting for slight disturbances which may indicate the presence of prey. If one pricks the surface near one of these animals with a pin it will turn towards the ripple, but after many such pricks it ceases to respond and takes a long time to recover. Even 24 hours later a series of tests will not give as many positive responses as in a fresh animal (Figure 4.6*b*). Nevertheless, there are a number of reasons for thinking that this is not action specific exhaustion. One is that an animal which has become fatigued by a series of 300 pricks on the water surface to the left of its head will start to respond strongly once more when the stimulation is moved to the equivalent position on the right side. Thus the exhaustion is specific to the location of the response and so does not apply to all responding. Another point is that recovery only occurs if stimulation ceases (Figure 4.6*c*). An exhausted animal will start to respond again after a few minutes if the water surface suddenly starts to be pricked again. But, if the stimulation has continued during these few minutes, recovery will not occur. In other words, it is not that a period without the action is required for recovery, but that the animal needs to be free of the stimulus for some time. This is stimulus specific waning of response or, in other words, habituation, and it is a very common phenomenon. We all know how one ceases to notice sounds, like that of a ventilator fan running continuously, but one jumps to attention when it stops or its tone changes.

Ceasing to respond need not therefore indicate that the *action* is exhausted. Furthermore, in some other cases, behaviour which is repeatedly elicited shows signs of becoming easier, rather than more difficult to produce: a process of sensitisation rather than exhaustion. The mating behaviour of male rats is an example which shows both processes (Figure 4.7). If a male that has not mated for some time is placed with a receptive female he will perform a series of mounts at the end of which ejaculation occurs; he will then groom for a while before a second series, and he may achieve three or four ejaculations in the course of an hour. In keeping with the idea that he is becoming exhausted is the fact that his refractory period, the time off he takes between mount series, increases from one copulation to the next. However,

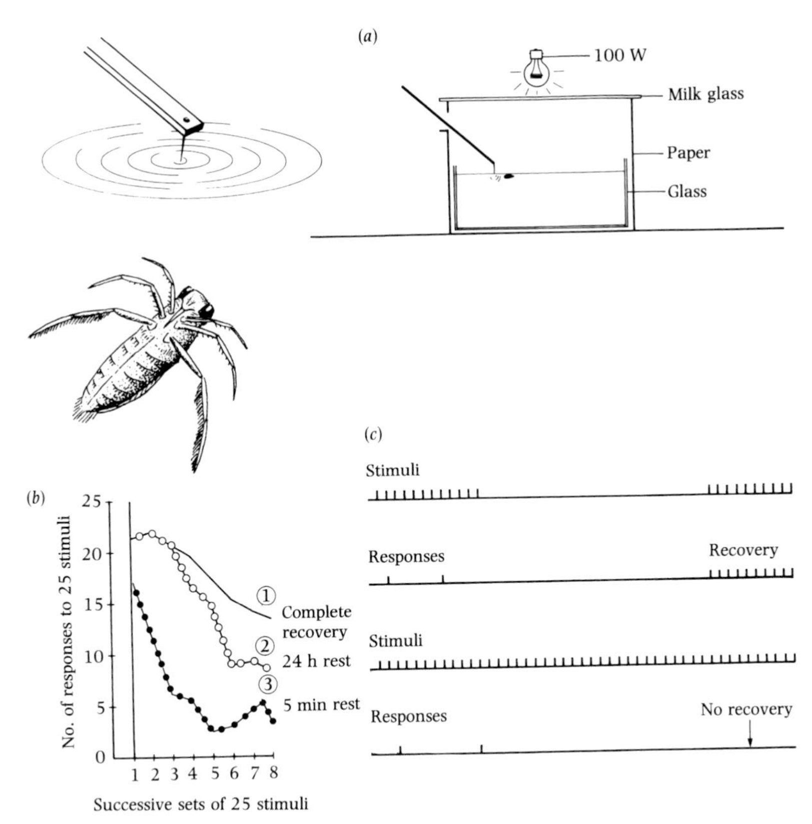

Figure 4.6. (*a*) The simple laboratory set-up used to study the response of the water boatman (*Notonecta*) to pin pricks on the surface of the water near it. (*b*) The number of responses in eight series of 25 trials in a fresh animal (1), one allowed to recover for 24 hours since last tested (2), and one given only 5 minutes to recover (3). The animal responds much less towards the end of the 200 trials, even if fresh, but shows some recovery when rested. However, even after 24 hours its recovery is not complete and it becomes exhausted more rapidly than it did the previous day. (*c*) Recovery will not occur if the stimulus continues, even if the animal does not respond to it. (After Wolda, H. 1961. *Arch. neerl. Zool.* **14**, 61–89).

the number of mounts he requires to reach ejaculation actually falls as the encounter proceeds: to begin with he needs a lot of stimulation; later on he needs much less. As far as this aspect of his behaviour is concerned, he is becoming sensitised rather than exhausted.

Examples such as these suggest that the changes in behaviour which animals show when repeatedly stimulated are far from simple and certainly cannot be thought of in terms of them running out of action specific energy.

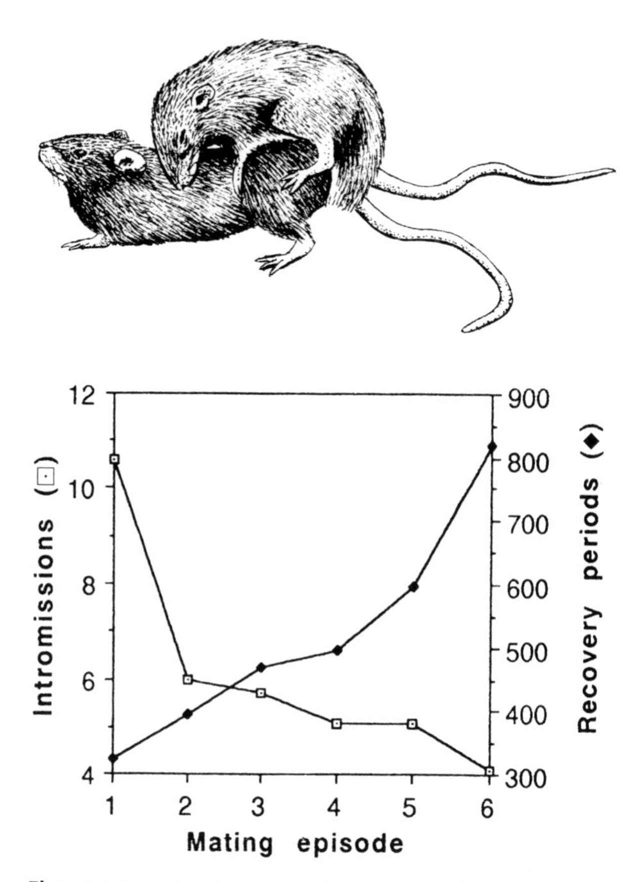

Figure 4.7. Graph to show the mating behaviour of a male rat placed with a female until he loses interest in copulating. On average he needs over 10 intromissions to achieve ejaculation the first time, and then starts mating again about 5 minutes later. After a few ejaculations only half as many intromissions are required, but it takes him twice as long to recover (Data from Beach, F.A. & Jordan, L. (1956). *Q. J. exp. Psychol.* **8**, 121–33.)

Sometimes they become more prone to show an activity rather than less; exhaustion of some aspects of behaviour may be rapid, of others slow; in some recovery depends on not showing the action, in others removal of the stimulus is required. Even in the male rat which has copulated to exhaustion it is not simply that his internal state has become incapable, as if he had run out of action specific energy. Give him a fresh female and he will start mounting again!

The conclusion then is that Lorenz's model only shows a very superficial similarity to the behaviour of animals as far as exhaustion is concerned. Their changes in responsiveness with repeated stimulation are far more complicated than the simple running out of energy that the model shows.

4.1.4 Internal and external factors

The last important point raised by Lorenz's model is that of the relationship between internal and external factors. The model suggests that the two add up to give behaviour, as both weights on the scale-pan and water in the tank can force the valve open, except that the tank must contain a minimum of water if the behaviour is to appear at all.

Just how internal and external factors relate to give behaviour is still a matter of contention and it may well be that some actions involve different relationships from others. It is certainly the case that behaviour patterns vary in the relative importance of internal and external factors in bringing them about. Feeding and drinking obviously need both: a rat will only drink if it is thirsty and if water is present; it will only eat if it is both hungry and presented with food. But consider two other examples: an exploring mouse and a singing bird. Mice kept in small cages explore rather little, but if they are suddenly presented with a new area into which they can wander they will start to move around it, sniffing into corners and rearing up onto walls. The larger the area, the longer they will spend looking round it. Thus exploration is quiescent for months on end, and then suddenly it emerges when a new environment presents itself. It is therefore a behaviour pattern largely driven from the outside rather than fired from within. The song of birds is just the opposite of this. Most male songbirds sing provided that they are in breeding condition, and the critical factor here is usually the male sex hormone testosterone. They will sing as long as they have enough of it in their bloodstream. The outside world is less important, but it does have some influence. Males may only sing on certain song perches, and they may sing more when it is warm and sunny, but the urge to sing is mainly from within them. Thus song and exploration represent two behaviour patterns at the ends of a spectrum, one mainly dependent on inside stimuli and the other on outside ones. Most other behaviour patterns, like feeding, grooming or sexual behaviour, lie somewhere in between these extremes and rely on both types of influence to a varying extent.

Regardless of which are the more important, just how do internal and external factors interact to give behaviour? Do they add together as Lorenz's model suggests? Once again it seems possible that the relationship varies between different types of behaviour. Addition is certainly a possibility,

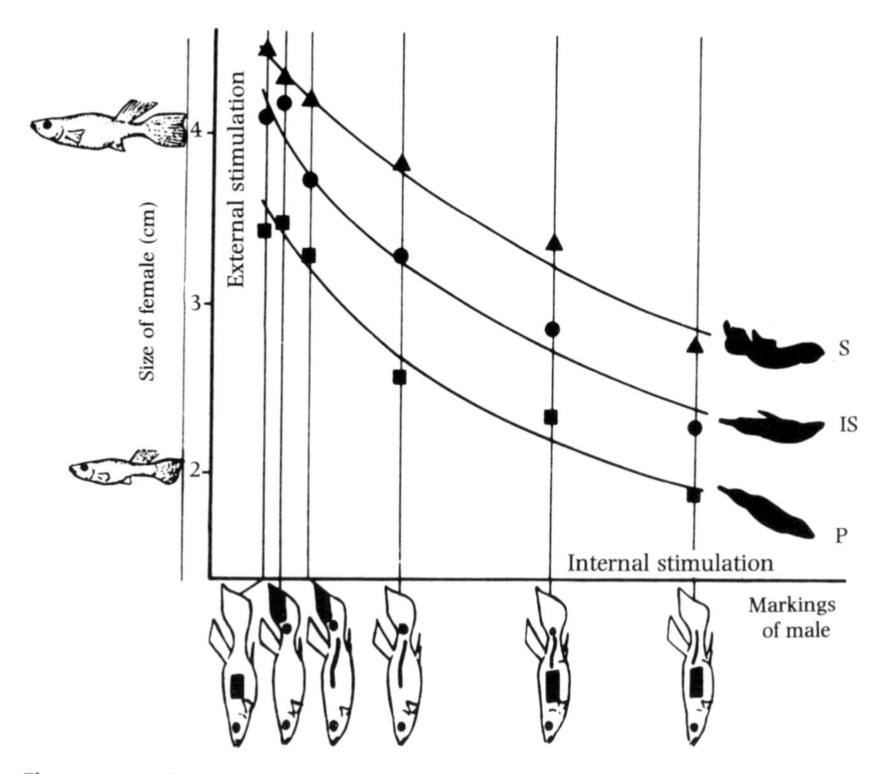

Figure 4.8. Three displays shown by male guppies are posturing (P), intention sigmoid (IS) and sigmoid (S). The graph shows the size of female needed to elicit these from males with various different markings. These provide a guide to the male's internal state, those on the right being more highly motivated that those on the left. (After Baerends, G.P., Brouwer, R. & Waterbolk, H.T., 1955. *Behaviour* **8**, 249–334.)

though it is an unlikely one because it suggests that behaviour may appear with no outside stimulus as long as the internal factors are high and, as discussed above, such vacuum activities are certainly not common. It also suggests that a high level of external stimulation might lead an animal to show the activity even if internal factors are very low. There is even less evidence for such an effect. Nevertheless it is extremely difficult to tell how internal and external factors do relate to one another. The problem is to decide 'how much' of each there is. The courtship of guppies will illustrate this point (Figure 4.8).

Male guppies court large females more vigorously than small ones. The male also indicates his internal state, as do many fish, by changing his colour patterns to become more brilliant when he is more highly motivated. Thus the colouration of the male gives a measure of his internal state and the size of the female is an indication of how good a stimulus she is. Plotting one against the other shows that they are clearly related. Males that are not well motivated will only display to females that are very large. On the other hand, males that are exceptionally brightly coloured will even display to very small partners.

This shows that internal and external factors interact to give rise to behaviour, but it does not tell us how. The size of a female could be assessed by her length, the area she projects onto the retina of the male, or many other measures. Assessing the colour patterns of males is even more difficult. Clearly some of them are brighter than others, so they can be put in an order, but there is no way that their distance apart can be accurately determined to give a scale of motivation. Thus, even after careful study showing how internal and external factors relate to one another in guppy courtship, it is not possible to say whether they add together, multiply, or have some more complex relationship.

This consideration of Lorenz's model of motivation shows just how far it has proved inadequate as a result of the many years of research which have elapsed since it was originally proposed. It was at that time a reasonable hypothesis, and has proved since to be a fruitful one. But most of its features have been found to be only crude approximations to the way that behaviour is really motivated. Two flaws stand out perhaps more than others. First, it has no role for feedback from the consequences of behaviour, an influence now realised to be extremely important for most behaviour. Second, it presents a picture of motivation which is the same for all behaviour patterns. Later research has shown just how different many of these are from one another and how dangerous it is to assume that the organisation of one is similar to that of another. Feeding, grooming, mating, drinking, fighting, singing and exploration all depend on their own particular internal and external causes and, furthermore, these may well differ from one species to another.

4.2 Deciding what to do

Animals generally only do one thing at a time, yet they often have the need to perform several. For example, a caged zebra finch waking up after a twelve-hour night must have a long list of priorities. These birds normally eat and drink about every half-hour, sing and fly around their cage for a period after this, and groom and rest till the next meal is due (Figure 4.9). Most of their grooming is brief, but every two hours or so they show a long bout which deals with all the areas of their body. At night they groom a little and rest a lot, but none of their other activities are performed during the hours of darkness. Not

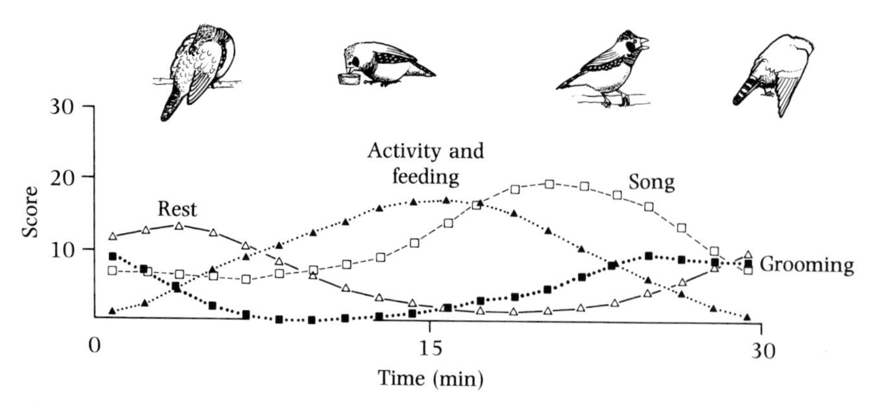

Figure 4.9. The short-term cycle of behaviour shown by a male zebra finch caged on his own. The illustration is based on observation of 12 cycles shown by a single bird synchronised at 30 minutes long by a cycle of changing light intensity. In each cycle the bird moves from resting to an active period during which it feeds, through a peak of singing to grooming, and then back to resting once more. (After Slater, P.J.B. & Wood, A.M. 1977. *Anim. Behav.* **25**, 736–46.)

surprisingly, then, they are rather busy in the first half-hour of the morning. Typically, they will stretch and ruffle their feathers when the lights come on, and then move over to their food for a long meal, perhaps followed by a drink. Some flying around and singing will follow this before they settle down for the first long grooming bout of the day.

Despite their need to do several things, these birds show a well organised sequence of behaviour. They do not rush around doing a little bit of one thing and a little bit of another, nor do they try to sing and groom at the same time. The sequence shown by different birds is similar and suggests that they all have the same priorities. The question of this section is how they decide between them, and it is a difficult question to which there is as yet no clear answer. What is clear is that the animal must make decisions and that to do so it must have some means of weighing up the internal and external factors relevant to different activities against each other.

The realisation that some sort of weighing up of options must go on within animals was one reason why psychologists postulated the existence of internal 'drives' which, like Lorenz's action specific energy, were thought of as powering behaviour and were also seen as being measured against each other, the strongest being the one that was expressed. A similar idea was developed by early ethologists, though they tended to use the word 'instinct' rather than drive. Animals were thought of as possessing a small number of instincts,

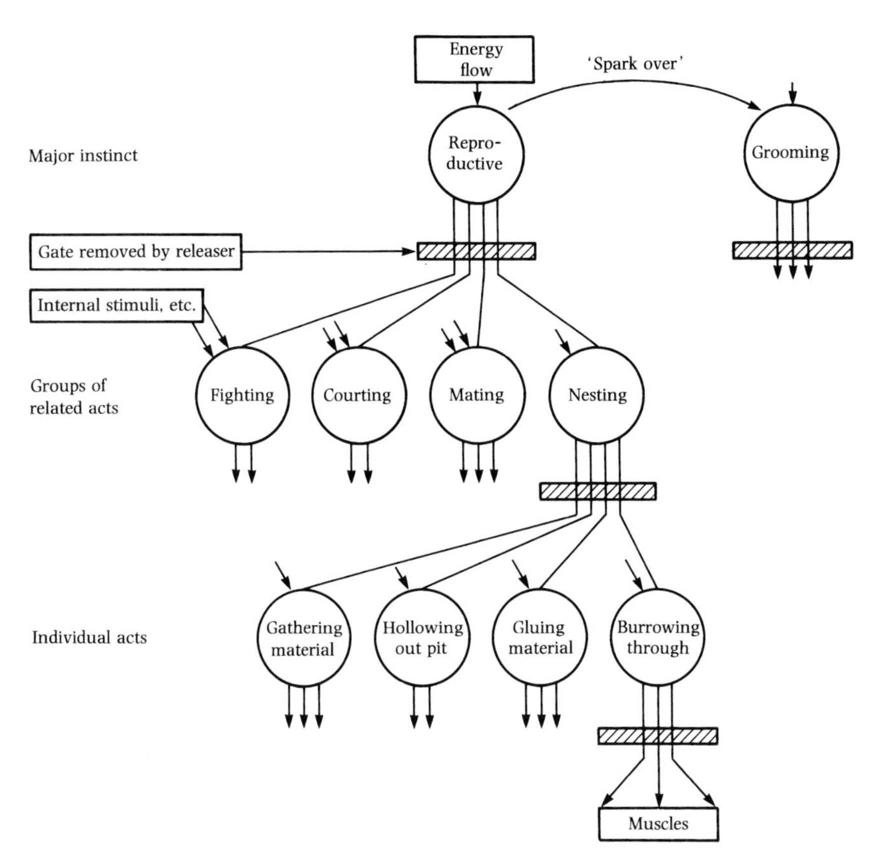

Figure 4.10. The essence of Tinbergen's hierarchical model was that a centre in the brain controlled each major 'instinct'. Energy from this flowed down to a series of lower centres when a gate was opened by the presence of appropriate releasers. In his original scheme Tinbergen saw the hierarchy being extended right down to the level of muscle units. Though the hierarchies of different instincts were separate, lack of releasers for one could lead energy from it to 'spark over' to another and so cause a displacement activity to occur. Thus, as in the example shown here, if reproductive activities were thwarted, grooming might appear.

such as those for feeding, aggression and reproduction, and it was they that competed with each other. The instincts themselves controlled a number of behaviour patterns: for example, the reproductive instinct led to various activities concerned with nest building, courtship and parental behaviour. Niko Tinbergen developed a model suggesting that each instinct might be organised in a hierarchy (Figure 4.10). Energy flowed downwards within this from the controlling centre at the top to centres responsible for groups of related activities and then further down to those concerned with individual actions. At each level the appropriate releasers had to be present, these acting to open gates through which the energy could flow to the level beneath and so on to give behaviour.

This is a ingenious idea, and unlike the 'as if' model Lorenz proposed, Tinbergen based it on the ideas of physiologists and he hoped it might have some neurophysiological reality, with various centres in the brain devoted to different instincts. Unfortunately it turns out not to be that simple. As mentioned before, the nervous system does not store and use up energy as these models suggest. Nor, unfortunately, is it neatly compartmentalised into centres and pathways with clear and distinct behavioural functions. Furthermore, though to some extent behavioural systems, like those controlling feeding, drinking or sexual behaviour, can be thought of as distinct, the factors affecting them overlap and they often influence each other. For example, the hormone oestrogen is an important internal factor making female rats receptive to the male, it also makes them very active so that they are more likely to come across a male, and it makes them less interested in food so that, when receptive, they spend less time eating and more looking for mates. This hormone therefore influences systems concerned with feeding, with activity and with sexual behaviour.

Rather than thinking of animals as having a number of discrete drives it is usual nowadays to think of them as having overlapping behavioural systems, each of them responsible for a group of related behaviour patterns (Figure 4.11). Actions with similar functions, like all the different movements involved in grooming or those concerned with nest building, tend to occur in close association, so it is reasonable to think of them as depending on a common system or, to use a more precise expression, as sharing causal factors. A causal factor might be a hormone or a particular level of food deprivation within the animal, or it might be a relevant outside stimulus, like the presence of irritation on its coat or of a member of the opposite sex. On this view of behaviour, whether or not a particular activity is shown depends on the combined level of its causal factors relative to those of all other possible actions.

For several reasons this is a more satisfying way of describing the motivation of behaviour than simply to consider it to be due to a series of drives. One of these is that the drive idea is only useful if all aspects of the behaviour can be thought of as affected by the drive in the same way. Careful studies of motivation run counter to this. Most obviously, a causal factor may raise some aspects of behaviour and lower others, whereas if it was enhancing drive, all aspects of the behaviour should be increased. Thus a very hungry animal may chew less and swallow more, and a bird building a nest actively

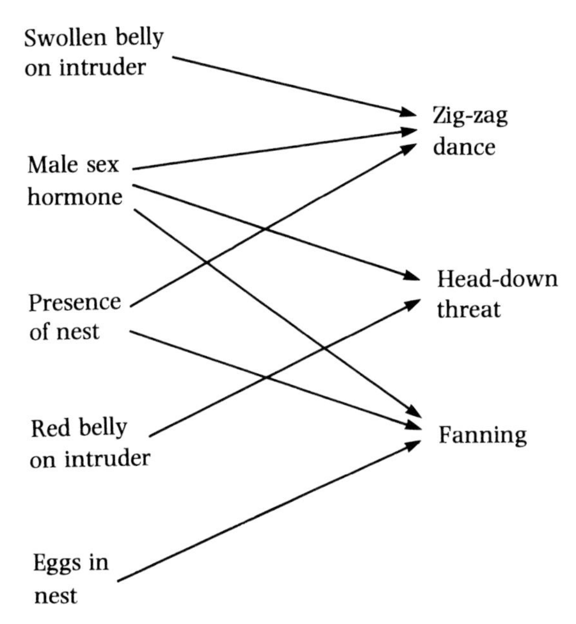

Figure 4.11. Current views of motivation do not see animals as endowed with a series of 'drives' or 'instincts'. Instead, different aspects of behaviour are thought of as influenced to varying degrees by numerous internal and external factors. Here some possible stimulating effects of five different causal factors on three male stickleback behaviour patterns are shown.

may fly to and fro with material at a great rate, but spend less time than usual carefully selecting and gathering material or in weaving it into the structure.

Another reason why viewing behaviour as dependent on a variety of internal and external causal factors is useful is that it helps to account for why behaviour patterns occur in association with one another to varying degrees. A grooming animal will wash its face, scratch its flank and shake its body, and these very different actions tend to occur close together rather than in association with feeding or mating movements. The most likely reason for this is that grooming actions share causal factors with each other, such as activity in a particular part of the brain or irritation of the coat. Alternatively behaviour patterns may be associated because one causes another. The fact that birds wipe their beaks after drinking comes into this category: drinking makes the beak wet and so produces a stimulus which is only removed by wiping. Thus, as well as behaviour patterns sharing causal factors and thereby appearing close together, they can be linked because one acts as a causal factor for another.

The way in which behaviour patterns may compete for expression has been most extensively studied in the case of feeding and drinking, because the causal factors for these two acts can easily be manipulated by depriving animals of food and water for different periods of time. An animal deprived of water tends not to eat as much (doubtless dryness in the mouth makes it difficult to chew and swallow), so the two actions are not totally independent of each other. But just how hungry and thirsty a deprived animal is can be established by seeing how much it eats and drinks when given food and water. We can then say it was 10 gm hungry and 1 gm thirsty if these are the amounts it consumes.

The most detailed experiments on the relationship between feeding and drinking have been those on doves by David McFarland. The doves he studied were placed in the 'Skinner boxes' that psychologists often use to study learning, in which they could peck one key to receive water and another to receive food. He found that a dove deprived of both food and water would alternate eating and drinking until it had satisfied both requirements. Its food and water needs, and how these changed as it ate and drank, could be neatly illustrated in the form of 'state space' plots such as those in Figure 4.12. If the bird was more hungry than thirsty it would start by eating (as in Figure 4.12*b*), and the amount that it ate could be traced by a line on the graph. After eating for a little it would then become more thirsty than hungry and would switch to drinking, the line on the graph moving so that the water deficit was shown as decreasing while that for food remained the same. Finally, after several drinks and meals, the line would have zig-zagged down to zero on both axes, indicating that the animal was satiated.

On this graph we can assume that every time the animal switches from one action to the other it is moving to that which has greater causal factors. We can draw a diagonal line which gives a rough impression of the boundary between feeding having priority and drinking doing so (Figure 4.12*c*). All switches in the drinking segment are from feeding to drinking, and those in the eating one are in the other direction. A feature which may seem curious here is that the animal goes well over the line in each direction before it switches rather than doing so immediately. If it did the latter it would end up 'dithering': after an initial large meal or drink to take it over the boundary line, it would oscillate rapidly from one activity to the other, taking tiny amounts of each in turn until eventually both needs were satisfied (Figure 4.12d). This would obviously be a badly organised way of doing things, but what makes sure that it does not happen? Several factors are probably involved, but one is likely to be especially important. This is that an animal that has just taken an item of food is likely to be looking at a food dish rather than at stimuli appropriate to any other action. As the sight of food is a causal factor for

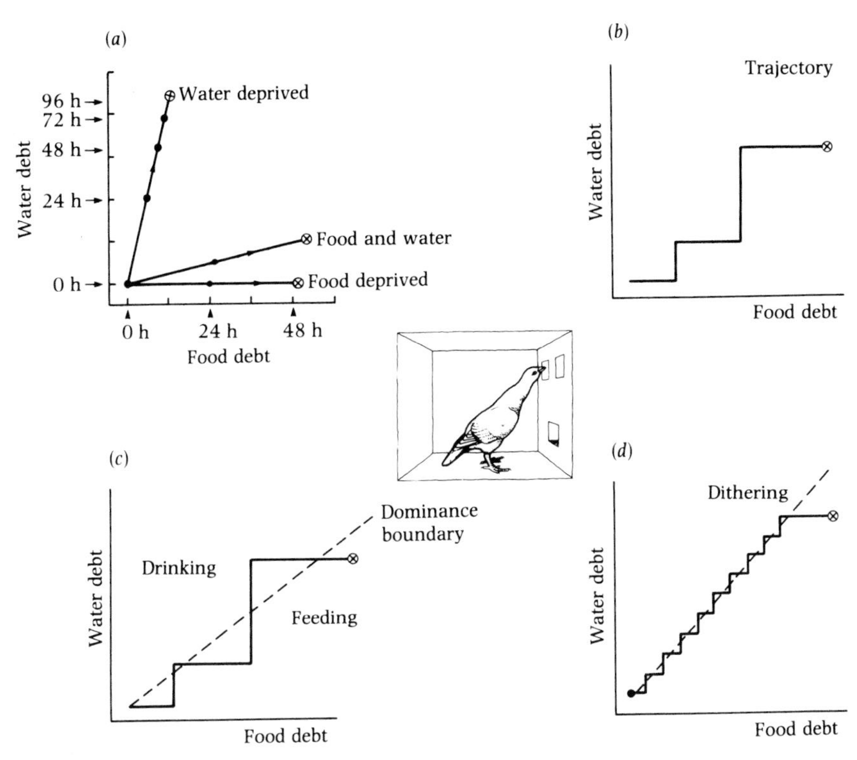

Figure 4.12. The way in which a deprived Barbary dove replenishes its reserves when placed in a Skinner box where it can peck keys to earn food and water. (*a*) The food and water debt of the bird can be represented by a point in a 'state space'. Here, three such series of points are shown, for birds deprived of food alone, food and water, and water alone to varying degrees. Birds eat less when deprived of water, so the points for this situation show a small food debt as well. (*b*) A bird deprived of both food and water eats and drinks alternately, so tracing a trajectory in its state space that takes it down to the point where it has cancelled both debts. (*c*) The boundary drawn to separate the area where the animal starts to feed (food dominant) from that where it starts to drink (water dominant). The animal tends to feed until it is well into the water dominant area and vice versa rather than dithering between the two, as shown in (*d*). (Adapted in part from McFarland, D.J. 1971. *Feedback Mechanisms in Animal Behaviour.* Academic Press, London.)

feeding, and that of water for drinking, this will mean that the animal is likely to continue with what it is doing rather than switching to another behaviour for which internal factors may well be high, but external ones are absent.

All this may seem quite complicated enough, but in the real world each behavioural system may itself have several dimensions. For example, animals may not just need 'food', but a correct balance between various different

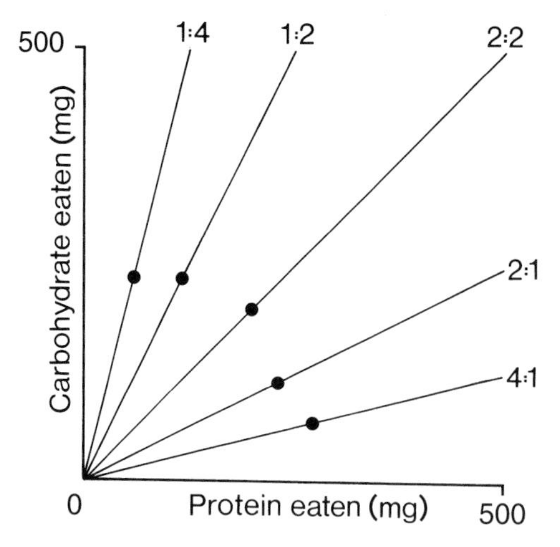

Figure 4.13. Locust instars need to achieve a balance of protein and carbohydrate in their diet. If presented with a food giving them the right balance they stop eating after their target is achieved. However, they cannot do this if the mix is wrong. This diagram gives five examples where they have been given different mixes, in the proportions shown by the ratios at the end of each spoke. As they eat, they are carried along the spoke that represents the proportions in the diet they have been given. The black dot on each spoke is the mean intake of 8 locusts on that diet. Over a large number of such tests they have been found to cease eating at the closest possible distance to the target. (After Raubenheimer, D. & Simpson, S.J., 1992. The geometry of compensatory feeding in the locust. *Anim. Behav.* **45**, 953–64.)

nutrients. Food may come in packages (a seed or an earthworm, for example) that do not present this ideal balance. Using the state space approach, Steve Simpson and his colleagues have examined how locusts cope with this problem when it comes to balancing their intake of protein and carbohydrate. Given a single source of food with a mix between the two which is not ideal, locusts cannot reach their optimum target, but they do home in on the point closest to it in the state space between the two (Figure 4.13). If given a choice of two foods, each with different mixes of nutrients, they can, however, move between the two different trajectories that these offer, and so get closer to their target. This is exactly what they do. They take some meals of one food and some of the other, their 'taste' for a particular food presumably depending on their requirement for what it provides at the time a meal is due to start. This allows them to meet their nutritional targets despite wide differences between the foods on offer.

In these examples the state space plots used express the relationship between two activities. In David McFarland's studies these were feeding and

Figure 4.14. Three classic examples of displacement activities: ground pecking in a fighting cock (*a*); sand-digging in a three-spined stickleback (*b*); wing preening in a courting garganey drake (*c*). (After Tinbergen, N., 1952. *Q. rev. Biol.* **27**, 1–32.)

drinking, but a third dimension might be added so that the animal's tendency to groom was also mapped in the hope that one might be able to predict when it would take a break from feeding and drinking to clean itself. Ultimately, one might hope to achieve mapping of all behaviour patterns onto each other in a 'multidimensional state space' but, as well as it being difficult to imagine such a complicated diagram, it is not easy to see what currency they could all be expressed in. There is certainly no such thing as 'grams' of grooming or of tendency to fight. So the problem of how the nervous system weighs up different actions against each other remains as real as ever.

4.3 Displacement activities

Watching a pair of courting birds, the observer might be surprised to see one of them break off to preen briefly and hurriedly before carrying on with its courtship. Similarly, a fish in the middle of threatening a rival may suddenly swim down and start digging in the sand as if switching to nest building, or a cock might interrupt fighting to take a few pecks at food (Figure 4.14). Such actions seemed irrelevant to those who saw them, and they often shared other features such as incompleteness or a hurried and frantic appearance, so they came to be grouped together and labelled as displacement activities. It was suggested that they arose when the animal was unable to carry on with more relevant behaviour because it was in a conflict as to what to do or was thwarted from achieving its aims. Thus the fighting gull might be in a conflict between attack and escape and so unable to do either; a courting male fish might have his sexual approaches thwarted by an unreceptive female and so, again, be unable to show the behaviour most relevant to the moment. In line with his hierarchical model of motivation, Tinbergen suggested that the thwarted energy from the reproductive instinct 'sparked over' and motivated behaviour down a different channel, such as that concerned with grooming (see Figure 4.10).

Ideas on displacement activities have changed enormously since they were thought to be due to sparking over. An important finding was that the causal factors which affect the behaviour in its normal context also affected it when it appeared as a displacement activity. A courting male stickleback might break off to fan at the nest, a behaviour which serves to aerate his eggs but which is irrelevant when there are no eggs there yet. Nevertheless, in its courtship context as well as when it appears later, the fanning is enhanced by carbon dioxide in the water. In the same way a bird which grooms when in conflict shows more of this grooming when its feathers are wet. Results such as these show that displacement activities are motivated in the usual way: their irrelevance lies only in their context.

The behaviour patterns labelled as displacement activities are undoubtedly a rag-bag of different actions, differently caused, the main thing they have in common being that someone studying the animals thought they were out of place. Conflict or thwarting may certainly be one reason why they arise: the animal, unable to carry out actions which are highest in priority, moves instead to ones which are next in line. This process is called disinhibition, and it is probably an important reason why displacement activities appear. It is perhaps not surprising that very often the displacement activities that have been described are grooming movements, because the animal carries around the stimuli for grooming with it so that, unlike food or water, they are always there on the outside of its body. Thus, if disinhibition occurs, grooming is very likely to follow. However, another reason why grooming often occurs out of context may be much simpler: the vigorous actions of fighting or courtship that the animal has been showing may lead its fur or feathers to be dishevelled and so actively enhance the stimuli for grooming till it breaks off for a quick preen to get rid of the itch.

In the days when behaviour was thought to be motivated by instincts or drives, it was these that were said to be in conflict with each other within the animal, or to be thwarted by a barrier to its behaviour on the outside. Thus conflict between the drives of aggression and of escape, or thwarting of the sex drive, were commonly referred to. However, these ideas were very vague: the drives themselves were hypothetical, so the conflict between them was doubly so. The theory was also virtually untestable: if everything was due either to its own drive or to a conflict between other drives, it was hard to think of any observation or experiment that could disprove it. Thus, as drive theories have fallen from favour, so has the idea of conflict between drives. But this does not mean that conflict or thwarting is not a real phenomenon and an important cause of switches in behaviour. Thwarting can easily be arranged by placing a perspex screen between a hungry animal and its accustomed food dish. Similarly, an approach–avoidance conflict can be set up by placing a frightening object beside a food dish. In such circumstances a whole array of behaviour patterns may be disinhibited because the animal is unable to perform those that are its top priority.

4.4 Motivation and animal welfare

Recent years have seen a growing concern over our responsibilities towards the animals that we keep as pets, in laboratories, in farms or in zoos. Are we giving them what they need? If not, are they being led to suffer as a result? Can we assess their requirements? Should we provide them with better conditions?

These are important issues, but not easy ones. It might seem obvious that one should try to keep animals in conditions as close as possible to those in the wild that they are adapted to, but this may be economically out of the question. Furthermore, the wild can be a pretty nasty place, where the lot of many animals is to die of starvation or be eaten by a predator. This said, there is no doubt that some captive housing conditions are very bare and cramped, and these can lead to abnormal behaviour, such as tail biting in pigs and feather pecking in chickens. Stereotypies, like the endless pacing up and down shown by some zoo animals in small cages, have also been cited as signs of distress, though they could equally be ways in which the animals have learnt to cope with their situation, and there remains some controversy about them.

As well as giving animals more space, efforts have often been made to make life more varied and interesting for them. Thus many zoos now provide enriched environments to give their animals things to occupy themselves with. One of their difficulties is that they have a lot of 'spare time': while it might take them several hours per day to collect enough food in the wild, in a zoo or farm this may be presented in a dish and wolfed down in seconds. This may not be the best thing for them. Interestingly, several studies have found that animals given free piles of food at one end of their cage and a bar that they could press to earn it at the other would spend quite a bit of their time pressing the bar.

Can we assess what animals really need? One approach that has had some success in this is adopted from economics, and this is to examine elasticity of demand. A hungry pig will press a bar to obtain a pellet of food, but will it press the bar five times, or 10 times, or 30 times for each pellet? According to

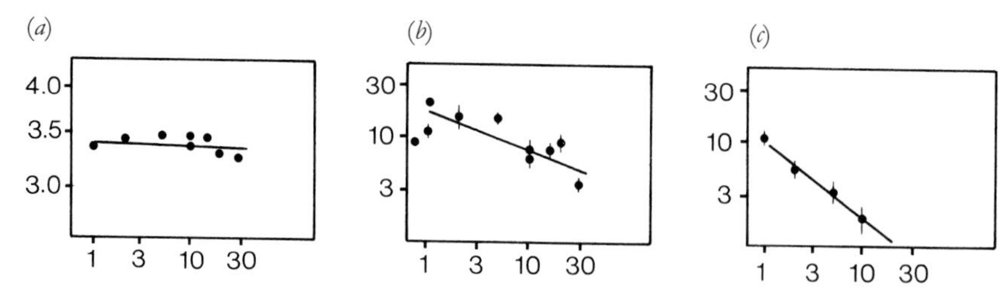

Figure 4.15. Bar pressing by a pig to obtain (*a*) food, (*b*) a view of a cage containing another pig, (*c*) a view of an empty cage. The number of bar presses required to obtain a reward rises from 1 to 30 along the X axis. This makes little difference to a hungry pig, which obtains almost the same number of rewards (Y axis, the log of the weight of food eaten in (*a*) and the number of rewards, also on a log scale, in (*b*) and (*c*)) regardless of the number of presses required. But the number of rewards obtained declines with the number of presses where social contact is the reward (*b*), and even more so where the reward is merely the door opening (*c*). (After Matthews, L.R. & Ladewig, J., 1993. *Anim. Behav.* **47**, 713–19.)

a study by Lindsay Matthews and Jan Ladewig, the answer is, yes: if it is hungry the 'price' of each food pellet can be raised substantially and it will continue to work (Figure 4.15). In jargon terms, the demand for food is 'inelastic', and the animals must have it. On the other hand, if the reward is the opening of a door behind which is another pig with which it can have some social contact, its interest in doing so falls off rapidly as the number of presses required is raised. The demand for social contact is much more elastic than that for food, and even more so when the opening door reveals only an empty cage.

Thanks to studies such as these, we are beginning to have a much clearer idea of what animals need and so how we should treat those for which we are responsible. This subject is an important way in which expertise in animal behaviour, and particularly an understanding of motivation, can be applied to practical problems.

4.5 Conclusion

Motivation theories have come a long way since the original ideas put forward by Lorenz and Tinbergen almost 50 years ago, yet the topic remains a difficult one. The study of individual systems, like those controlling eating, sexual behaviour or exploration, has advanced mostly in the cases of those in which the internal state of animals can be altered physiologically to discover just what causal factors do operate from within. The influence of hormones on sexual and aggressive behaviour, and of nutrients and ion balances on eating and drinking, have been studied extensively, as have the neural pathways involved in all these actions. Nevertheless, behavioural work aimed at understanding how internal and external factors interact so that behaviour may be predicted with greater certainty, remains useful and important as a complement to these physiological studies. The complexities of the interactions between behaviour patterns and of the decision making processes which lead animals to select which they will perform are even greater, and here studying physiology will be of little help. Searching for the rules which govern switches from one behaviour pattern to another is the only option: progress has been made, but the problem remains a daunting one.

5

Development

The problem of how behaviour develops has been among the most contentious in the whole field of animal behaviour. The last few chapters have illustrated the stress laid by the early ethologists on words such as 'innate', 'inherited' and 'instinctive' to describe the behaviour patterns they studied. Thus the German word for fixed action pattern, *Erbkoordination*, means, literally translated, inherited coordination. The sensory routes through which these actions are elicited were termed innate releasing mechanisms, and the motivational systems controlling behaviour from within were often referred to as instincts. All these terms implied great rigidity in behaviour and its development, with no scope for the environment to have any influence, as if the action was utterly inevitable and its whole form and development simply a readout of genetic instructions.

On a view such as this, development was obviously rather an uninteresting process, the appropriate behaviour springing fully formed from the animal at the time that it was required without the need for complicated developmental processes involving interactions with the environment. But it was not a view that went uncontested and, in particular, it contrasted with that of the 'behaviorist' school of psychology in America which stressed the prime importance of learning in the development of behaviour. For a long time the main issue in behaviour development was this controversy between learning and instinct, so we must start here by considering these opposing views before discussing more recent work on exactly how behaviour does develop.

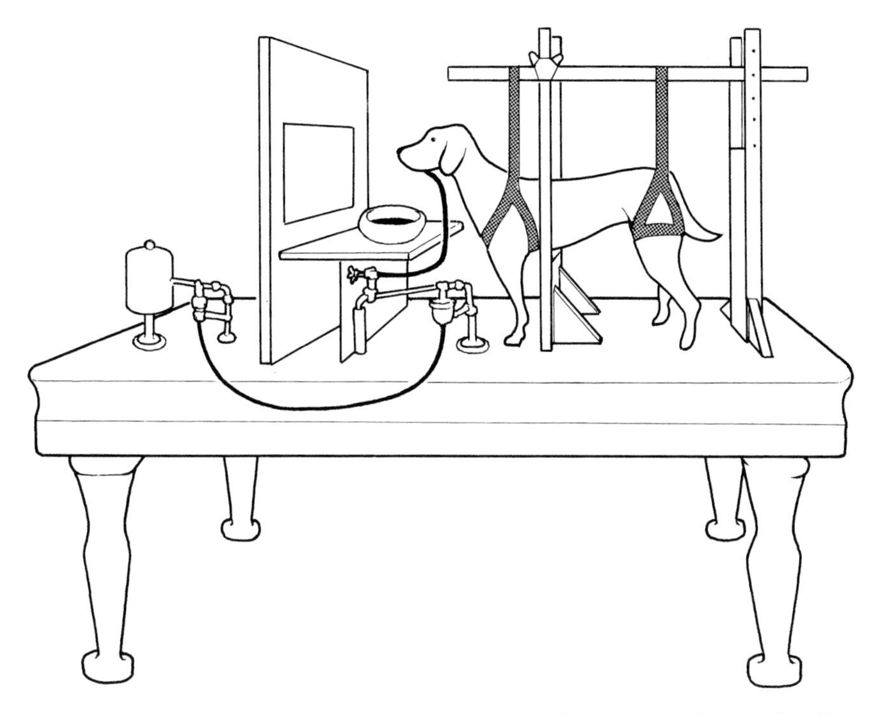

Figure 5.1. The dogs studied by Pavlov were held in harnesses with one of their salivary glands connected through a tube to a collecting bottle. He was able to present them with the stimuli he used, such as food and the ringing of a bell, from behind a screen, and then see how much saliva they had produced as a result.

5.1 Psychologists and learning

The major figure in the behaviorist school of psychology was J. B. Watson. He followed the English philosopher John Locke in believing that the innate mind was a '*tabula rasa*': a blank slate on which subsequent experience was written. Watson studied learning in animals such as rats and was impressed by the flexibility of their behaviour and by how learning could lead them to behave in an adaptive manner. He believed that they built up connections in their brains as a result of perceiving associations in the outside world, so they would associate a particular place with food and run through a maze to get there. His views were partly stimulated by the work of Ivan Pavlov, the Russian physiologist, who had studied conditioned reflexes in dogs. Pavlov showed how a new stimulus could be made to elicit a reflex as a result of the animal building up an association. In his most famous experiment (Figure 5.1), a dog was presented

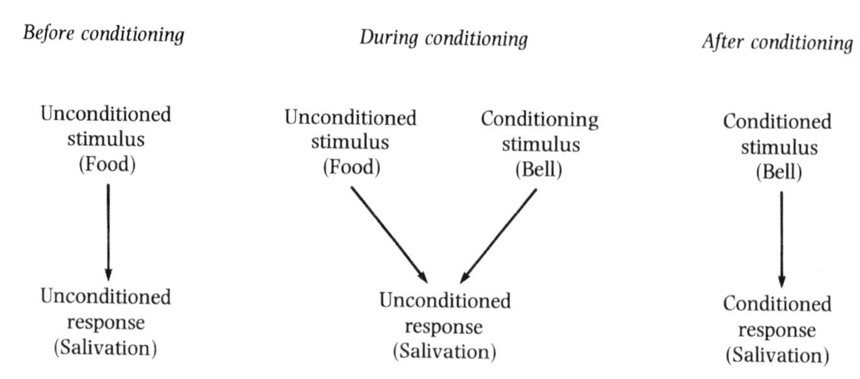

Figure 5.2. Conditioning involves the formation of new connections. Before it takes place a stimulus leads to a response. As a result of that stimulus being associated with a different one (the conditioning stimulus), the second stimulus comes to lead to the response. In the case studied by Pavlov, the bell came to lead to salivation without the need for food to be present.

with food, a stimulus which normally makes dogs produce saliva in readiness to eat, and at the same time a bell was rung. This normally has no effect on salivation. After a number of such presentations, the bell was suddenly sounded without any food being produced. The dog salivated. The animal had formed an association between the bell and the food so that both were now capable of producing the response. This sort of training is referred to as conditioning and the result, in this case salivation in response to a bell, as a conditioned reflex (Figure 5.2).

The main stress in the theories of Pavlov and Watson was not on reward: the dog did not need to eat the food to form the connection. Later ideas have suggested that a great deal of learning does rely on reward, the most notable exponent of this view being B. F. Skinner. Like Watson before him, Skinner believed that the great majority of behaviour, with the exception of simple reflexes, results from experience. But he stressed the way in which the responses of animals come to be linked to particular stimuli as a result of reward. Animals tend to repeat rewarded actions and reward can thus be used to alter their behaviour. The dove in the Skinner box referred to in the last chapter soon learns that pecking the right hand key gives it water while the left key yields food. Based on ideas such as these, elaborate theories of animal learning have been built up which stress the rôle of the environment and of experience in the development of behaviour rather than genetics and inheritance. At an extreme such psychologists have suggested that learning in all animals is subject to the same rules and that animals do not have predispositions about the sort of things they will learn but will build up any associations with equal ease provided that their sensory and motor systems allow them to do so. Obviously one would not attempt to train a sparrow to fly in the dark or a tortoise to walk on its hind legs.

The following famous assertion of Watson's illustrates the fervour of his belief that the environment rather than heredity was the major determinant of behaviour:

> Give me a dozen healthy infants, well-formed, and my own specified world to bring them up in and I'll guarantee to take any one at random and train him to become any type of specialist I might suggest – doctor, lawyer, artist, merchant-chief and, yes, even beggar-man and thief, regardless of his talents, penchants, tendencies, abilities, vocations, race of his ancestors.

He went on to admit that he was overstating his case, but insisted he was doing so as a counter to the sweeping claims of his opponents.

Both views such as this and those of the ethologists could not be right, but to some extent they existed because their adherents were studying different things. Psychologists were interested in learning and in behaviour patterns, like the lever pressing of a rat in a Skinner box, that could be modified by it so leading an animal to behave differently from other members of its species. By contrast, the ethologists were interested in more fixed behaviour, such as courtship displays, which were common to all members of a species. Why did they consider such behaviour to be 'innate' and how does the evidence they used hold up to detailed examination?

5.2 Ethologists and instinct

In 1828 a boy of about 16 was found in the market-place at Nuremberg in Germany. He could not speak and his behaviour was like that of a little child, so he was labelled 'the wild boy'. He came to be called Kaspar Hauser and, when he was able to communicate, he explained that he had been brought up entirely in isolation by a man who had kept him and cared for him all by himself in a hole. Kaspar Hauser may well have been a fraud, but he has given his name to the deprivation experiments which have often been used by ethologists in an attempt to understand whether or not the behaviour they were studying was innate. The idea was to deprive an animal of relevant aspects of experience and, if the behaviour appeared despite this, it could be regarded as inherited whereas, if it did not appear or developed abnormally, it could be assumed that the missing experience was important.

There are many striking examples of behaviour appearing apparently

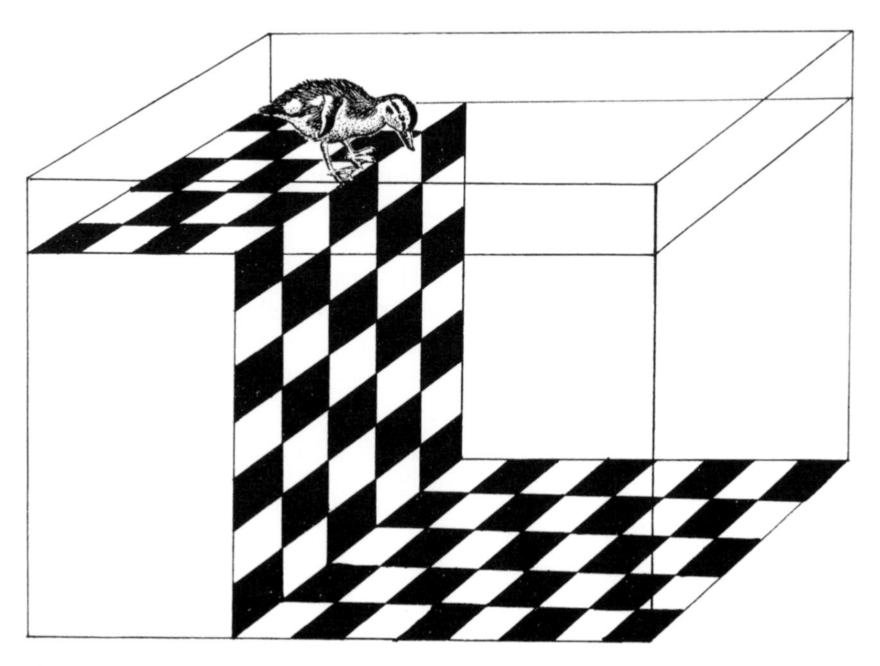

Figure 5.3 A visual cliff, as used by Kear in her studies of ducklings. The whole apparatus is covered by a sheet of glass, but the left side appears to be shallow while that on the right looks deep.

normally in animals whose experience is distinctly abnormal. A case in point was shown by experiments carried out by Janet Kear on ducklings which she had reared by hand from the egg. Most duck species nest on the ground, but some of them, such as the wood duck, do so in holes far up in trees from which the fledglings have to leap to the ground beneath. Fortunately they are light and fluffy when they do so and, as a result, they bounce and run off rather than suffering multiple fractures. Kear tested chicks of various species on a visual cliff such as that shown in Figure 5.3. In this apparatus the animal is placed in the centre and can move either to one side where there is a drop looking like a cliff beneath the glass or to the other where the floor is immediately under the glass so that it looks shallow. The behaviour of the ducklings was appropriate to their normal nesting place. Tree-nesters did not avoid the deep side of the cliff but, if they moved that way, they would leap as if casting themselves into space. On the other hand, the ground-nesters tended to move to the shallow side rather than the deep one, suggesting that they avoided heights. Interestingly, if they did move to the deep side their behaviour was quite different: they pushed off with both feet as they would when moving out from the edge of a pond!

Being hand-reared, these young birds had had no opportunity to learn the actions they showed from others or from earlier experience with ponds or with cliffs. Many ethologists would therefore have used this evidence to argue that the behaviour must be innate because it develops despite deprivation of opportunities for learning, and the main motivation for carrying out such experiments has often been to discover whether behaviour is 'innate' or 'learnt'. Sometimes, as with the ducklings, it develops normally even though the animal is reared in a very impoverished environment. Another good example here is the hoarding behaviour of squirrels whereby, even in captivity, they will bury nuts underground to form stores which they eat later when food is scarce. If such an animal is raised on a liquid diet, so that it never experiences nuts, with masses of this food available the whole time, so that it never needs to hoard, and on a bare floor so that digging is impossible, the first time it encounters nuts and earth it still digs a hole and buries them.

By contrast with these experiments, others have shown behaviour patterns to be radically altered unless particular experiences are available. If a young dog is reared in total isolation from all others, in a chamber in which it can be fed and cared for without contact with its human caretakers either (Figure 5.4), it turns into a strange animal, apparently careless of its own welfare. It will put its paw in a fire and singe it, and it will repeatedly approach an object that gives it an electric shock. This is quite unlike a normal puppy of the same age, which hastily withdraws from such painful experiences. Furthermore, the normal puppy will yelp when hurt, whereas the one that had been isolated behaves as if it did not even feel the pain. Clearly its deprivation has led to a drastic alteration in the way its behaviour developed. In another example, to which we will return later, a young male chaffinch reared out of earshot of all birds of its species has been found to develop a very simple and unstructured song quite unlike that of a normal adult. If he is deafened as well, so that he cannot even hear his own efforts at singing, the song he produces is even worse, being little more than a screech.

These Kaspar Hauser experiments vary enormously in what they actually deny the animals. The isolated chaffinch cannot copy song from other birds, but he can practise singing; if deafened he can still practise, but cannot hear the outcome. The hoarding squirrel has had experience of neither nuts nor earth so its deprivation is more extreme: it cannot copy from others, it cannot learn for itself, nor can it practise digging in earth, though it can carry out the movements concerned on the bare cage floor. In many cases it has proved very difficult to deny animals all the experiences one might think likely to be relevant. Finches with no hay that they can use for nest building will carry seed, lettuce, faeces and feathers to their nest site. Some bizarre behaviour results

Figure 5.4. Studies of behavioural development in dogs involved rearing puppies in isolation both from other dogs and from humans. They were kept in chambers with two doors and a partition so that one part could be cleaned out while they were in the other. As a result, they were isolated from all contact with the outside. (After Thompson, W.R. & Melzack, R., 1956. *Scient. Amer.* **194**, January.)

when the feathers used are still attached to themselves or to their mates. A bird may have to climb rather than fly to its nest site because it is holding its own wing in its beak as nest material; after carefully placing the wing in a corner of the nest it will then fly down, pick the wing up again and struggle back to the nest with it!

5.3 A false dichotomy

To ethologists such as Konrad Lorenz deprivation experiments were the main evidence used in deciding whether or not particular aspects of behaviour were innate. But the examples mentioned above point to a problem in interpreting these experiments: it is almost impossible to tell exactly what one has deprived the animal of. This difficulty was one of those pointed out forcefully by a leading critic of Lorenz's theories, the American psychologist Danny Lehrman. The points he raised deserve discussion here because they were very influential in leading to current views of behaviour development. They fall under three main headings.

5.3.1 Criticism of the deprivation experiment

Lehrman argued that deprivation could show where experience *was* important but not where it was not. Thus the chaffinch which sings abnormally after being isolated shows that hearing the song of others is essential for normal song development. But the bird that builds a normal nest the first time it encounters hay may have had many experiences, from grooming its own feathers to husking seeds, which could have contributed to its nest building capability. Sometimes the strangest and most unexpected experiences turn out to be important. Baby rats will not urinate for the first time unless their genital area is stimulated: normally this occurs because their mothers lick them soon after they are born, but if they are isolated before this occurs they will swell up until their bladders burst. Given the existence of such unlikely influences, it is rash to assume that a simple deprivation experiment has removed all the experiences that may be relevant.

5.3.2 Criticism of viewing learning as the only environmental influence

The environment has many influences on animal development to which it would not be appropriate to apply the word learning. The example of the baby rats can also be used to illustrate this point. Though the young rat is stimulated to start urinating, no one would suggest that it has learnt to do so. Many other environmental influences which affect behaviour have also been described. Kittens growing up in an enclosure covered with vertical stripes become incapable of seeing horizontal ones, and those reared with the opposite orientation show the reverse effect (Figure 5.5). In toads, the visual areas on the two sides of the brain are wired together so that cells seeing the same point in space become connected to each other as a result of their common experience rather than by some automatic program. As toads can reach very different sizes, this strategy of developing connections is a good one if they are to achieve the perfect binocular vision which is essential to them for prey capture.

These examples all concern environmental inputs which will, directly or indirectly, have an effect on animal behaviour but none of which would be

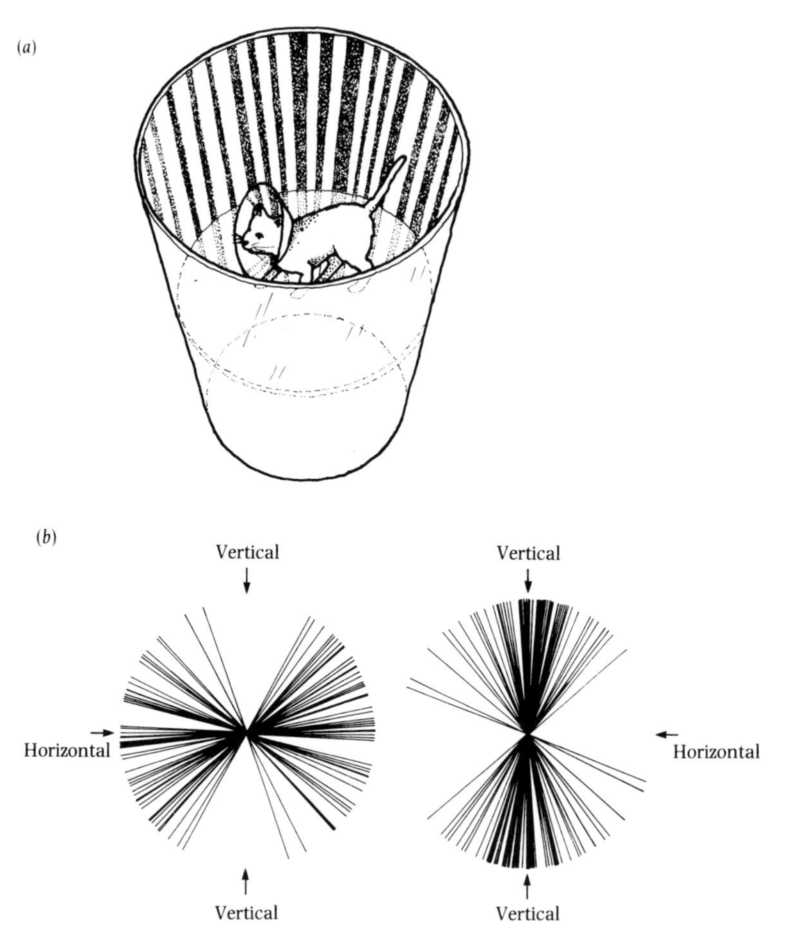

Figure 5.5. The kittens studied by Blakemore & Cooper (1970. *Nature* **228**, 477–8) obtained all their visual experience in tubes lined with either vertical or horizontal stripes (*a*). As a result the units in their visual cortex were found to be sensitive only to lines in the orientation they had experienced (*b*). Each line in (*b*) represents the preferred orientation of a cell in the visual cortex of one of the kittens, an animal reared with horizontal stripes on the left, with vertical on the right.

called learning. In some cases, like the failure of young rats to urinate unless they are licked, the effect is quite specific. In other instances, as where an abnormal environment leads to a visual deficit, a much more general influence on numerous different aspects of behaviour is bound to occur. Many such environmental effects are now known, and those such as nutritional deficits and oxygen shortages at particular stages of nervous system development are examples of ones known to have a profound and lasting effect on behaviour.

5.3.3 Criticism of viewing behaviour as either/or

The discovery that many unexpected factors can influence behaviour led ethologists to examine the development of behaviour much more carefully and to avoid making sweeping statements about how this or that behaviour pattern is 'innate' or 'inherited'. These words suggest that the behaviour is absolutely fixed and that it would develop exactly the same no matter what environment the animal found itself in. Konrad Lorenz clearly felt that many behaviour patterns were like this as he referred to them as being 'blueprinted in the genes'. But, as Lehrman pointed out, expressions like this give a false impression of behaviour development. By depriving the animal of certain influences, one can show that they are unnecessary for its behaviour to develop normally. But where should one stop? There are obviously some very specific effects one might predict as being important to a particular behaviour. For instance, experience with nest material or seeing another animal nest building might be two that could well be essential to normal nest building development. However, there is a continuum of possible influences from ones such as these to very unlikely ones, like experiences during feeding or grooming, which do after all, like nest building, involve manipulating things in the mouth. One cannot be sure that experiences such as these have no importance for nest building. Ultimately some environmental influence is bound to be important in the case of every behaviour pattern because behaviour does not exist in the genes and the genes have to have an environment in a particular temperature range, supplied with oxygen and with nutrients, and so on before the information they contain can be read out to give an animal.

5.4 The conflict resolved

As a result of such criticisms, an important point that the ethologists came to appreciate is that natural selection has only determined how development should take place in the *normal* environment of each species. Thus a behaviour pattern may appear extremely fixed and constant in all individuals of a species because their genes interact with that environment to ensure that this is the case. However, moved to another environment, different from any the species has encountered before, the result may be quite different. Natural selection does not ensure that the genes will interact with *any* environment to give the same result; it can only give the right outcome in environments in which it has had a chance to work. Hence domestic animals, or for that matter humans, may develop quite differently from the way they would have done in nature. This can lead to behavioural problems, but it is also a point of optimism: even if natural selection favoured individuals that were highly aggressive, this is no reason to think that we are irrevocably committed to being so. Given a different environment from that in which evolution took place, the same genes may produce individuals with a quite different mix of characteristics which could easily be much less aggressive.

On the other side of the dichotomy, the findings of ethologists have also had a marked impact on how psychologists view their subject. For, of course, it is equally sweeping to refer to a behaviour pattern as totally environmentally determined as if the animal's genes were of no significance whatsoever. As psychologists have studied learning in a wider diversity of animals they have found that different species have different capabilities and that they find some tasks easy to learn while others, which might seem similar at first sight, prove very difficult. Such results are often referred to as demonstrating 'constraints on learning'. They indicate that an animal's heredity has a profound impact on what tasks it is capable of learning to perform.

Most psychologists interested in learning still study rats and pigeons but, even within these species, some interesting constraints have been found. It is easy to train a rat to press a bar to obtain food but not to avoid an electric shock, mainly because the natural reaction of a frightened rat is to 'freeze' and this is incompatible with bar pressing. In some instances where training was successful the rat was found to be freezing on top of the bar rather than pressing it with a paw! In another experiment, shown in Figure 5.6, it was found easy to get rats to associate a taste with sickness and a sound with pain but hard to train them to make the opposite pairings. This result makes biological sense: in nature tastes are more likely to be a good indication of which food is bad for you than are sounds, whereas pain is more likely to be experienced alongside sounds such as those made by the approach of a predator. But results such as these were something of a surprise to those psychologists who were committed to the idea that an animal's heredity was of no significance and that it should therefore make various different associations with equal ease.

Ethologists and psychologists had a lot to learn from each other as far as behavioural development was concerned, but the controversy led to a genuine coming together of the two fields. Most would now agree with Donald Hebb's remark that both genes and environment are 100% important in the development of all behaviour: no genes, no behaviour; no environment, no behaviour. And, as we shall see in the rest of this chapter, some of the interactions between the two are fascinating and subtle. It certainly does not do justice to them to ask simple questions about whether behaviour is genetically or environmentally determined.

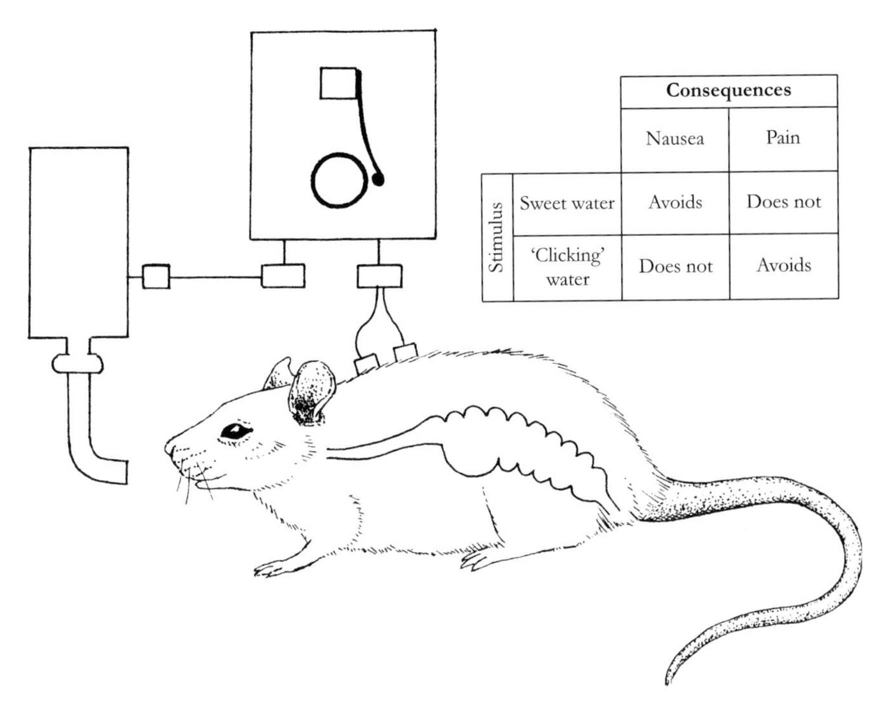

		Consequences	
		Nausea	Pain
Stimulus	Sweet water	Avoids	Does not
	'Clicking' water	Does not	Avoids

Figure 5.6. Rats that have drunk sweet water and become sick avoid that water again, and rats in which a clicking sound is followed by electric shock show defensive behaviour after they hear the click. But rats do not easily learn to associate a clicking sound when they drink with nausea, or sweet water with electric shock. (After Garcia, J., Clarke, J.C., & Hankins, W.G., 1973. In *Perspectives in Ethology*, ed. P.P.G. Bateson & P.H. Klopfer. Plenum Press, New York.)

This argument does not mean that carrying out deprivation experiments is a useless enterprise. Depriving an animal of particular experiences can indicate whether or not they are crucial for normal development. Furthermore, deprivation experiments have shown that some aspects of behaviour are much more flexible than others. Certain actions, like hoarding in squirrels, develop much the same even though the environment is changed substantially in various different ways. Others are altered by even minor manipulations. Thus some behaviour patterns are stable and some labile, and it is interesting to examine the reasons for these different developmental strategies. The consideration of a few case histories will enable us to do this, at the same time as illustrating how behaviour develops now that it is no longer adequate simply to label it as 'innate' or as 'learnt'.

5.5 Some case histories

5.5.1 Song development in birds

Bird song is the classic example of how both genes and environment have a crucial role to play in development. Since the pioneering work of W. H. Thorpe on chaffinches, many species have been studied and it has become clear both that learning plays an important role in all of them and also that there are constraints on what they are able to learn.

Thorpe was able to show that learning from others was involved in chaffinches by a series of experiments on hand-reared chicks. As in most other species, at least in temperate regions, only the males sing. Thorpe found that, if he raised young males in total isolation from all others, the song they produced was quite different from that of a normal adult. It was about the right length and in the correct frequency range; it was also split up into a series of notes as it should be. But these notes lacked the detailed structure found in wild birds, nor was the song split up into distinct phrases as it usually is. This suggested that song development requires some social influence. Later experiments in which young birds were played recordings of songs showed just how precise this influence was: many of them would learn the exact pattern of the recording they had heard (Figure 5.7). A remarkable feature here was that birds were able to copy precisely songs that they only heard in the first few weeks of life, yet they did not sing themselves until about 8 months old. They are thus able to store a memory of the sound within their brain and then match their own output to their recollection of it when they mature.

Young chaffinches normally learn only chaffinch song, though Thorpe found they could be trained to sing the song of a tree pipit, which is very similar to that of their own species. In general, however, the constraints on learning which birds have ensure that they only learn songs appropriate to the species to which they themselves belong. These constraints may be in their neural circuitry, the young bird hatching with a rough idea of the sounds that it should copy. The crude song of a bird reared in isolation gives some clues as to what this rough idea may be: the length, the frequency range and the breaking up into notes are all aspects of chaffinch song shared between normal birds and those reared in isolation. In other cases the constraints are more social, young birds only being prepared to learn from individuals with whom they have social interactions. Thus, in a number of species, it has been found that they will not copy from recordings, but will do so from a live tutor. In some cases this may occur when they are fledglings, but in others the main learning period is when they set up their territories and interact with

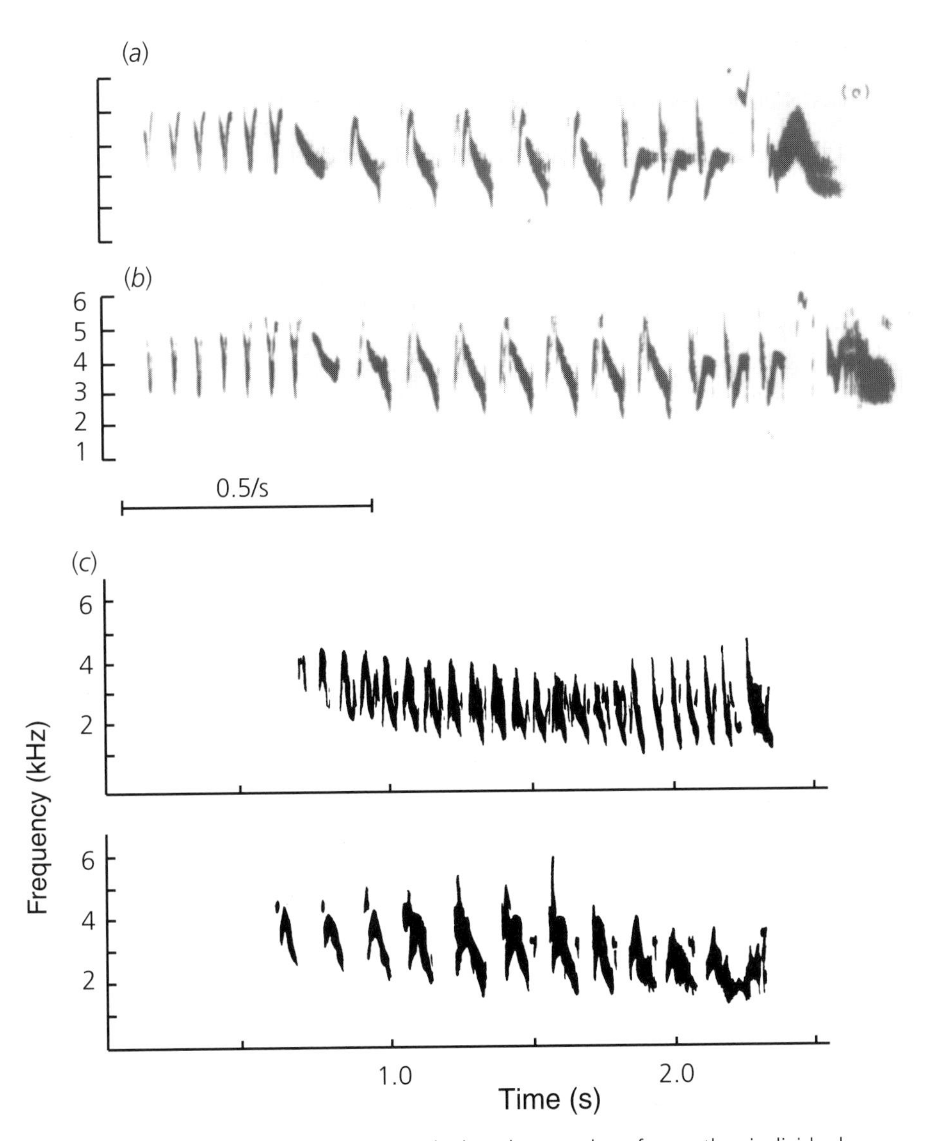

Figure 5.7. Birds often learn the sounds that they produce from other individuals very accurately. Just how well they do so can be assessed by examining the traces made by a sound spectrograph, as shown here. This machine produces a plot of frequency in kHz (1 kHz = 1000 cycles/second) against time, the darkness of the plot showing how much energy was present at that particular moment and pitch. (*a*) shows the song phrase of a chaffinch played to a young individual when it was starting to sing and (*b*) the copy that it produced. (After Slater, P.J.B. & Ince, S.A., 1982. *Ibis* **124**, 21–6). (*c*) shows the 'Kaspar Hauser' song of two young chaffinches reared in isolation. (After Thorpe, W. H., 1958. *Ibis* **100**, 535–70.)

neighbours for the first time, enabling them to match their neighbour's songs and so counter-sing with them. Whatever the nature of the learning rules in a particular species, there is no doubt that they are effective: it is very unusual to hear a wild bird singing a song which is not typical of its own species despite the many different songs which often occur in a small patch of woodland.

Not all birds show the same learning pattern as do chaffinches. There are some species which produce normal sounds even if deafened, so that they cannot hear their own efforts, far less copy those of others. The cooing of doves and the crowing of cocks are examples here. In other cases, such as parrots and hill mynahs, birds can be trained to copy a huge variety of sounds, though those they learn in the wild are usually more restricted. The amazing capability of mynahs has apparently arisen simply because birds in an area learn a small number of their calls from each other, males from males and females from females, and these calls are highly varied in structure. The ability to master them has led the birds, incidentally, to be capable of saying hullo, copying wolf-whistles and mimicking a wide variety of other sounds. The capacity for mimicry in some birds can be just as remarkable in the wild and shows that not all of them are as constrained as the chaffinch in what they will copy. The starling and the mockingbird are perhaps the best known examples but a less well known species, the marsh warbler, must take the prize. Françoise Dowsett-Lemaire found that males of this small migrant learn sounds from many other birds as they move down from Europe into Egypt and then on to East Africa in the first autumn of their lives. They then string them together into an amazingly varied song on their return north again the next spring (Figure 5.8). The average number of species that a male imitates is 76: the only sounds he seems to reject are the really deep ones which he cannot master because of his small size!

The timing of learning also varies a lot between species. In chaffinches it is restricted entirely to the first year of life, and adult males that have not learnt a normal song by then cannot be trained to produce one thereafter. The male sex hormone testosterone, which induces them to start singing, seems also to stop them learning more. Their learning thus occurs in a short 'sensitive period' early in life. Such sensitive periods are a common phenomenon in behaviour development, but they are not always found. While some birds may only learn as fledglings or as young adults or, as in the chaffinch, at both these times, other species carry on learning throughout life. An example here is the canary: males in this species modify their song from year to year regardless of their age.

Bird song, then, provides a fine range of examples of behaviour development, some tightly constrained and others highly labile. But, despite the

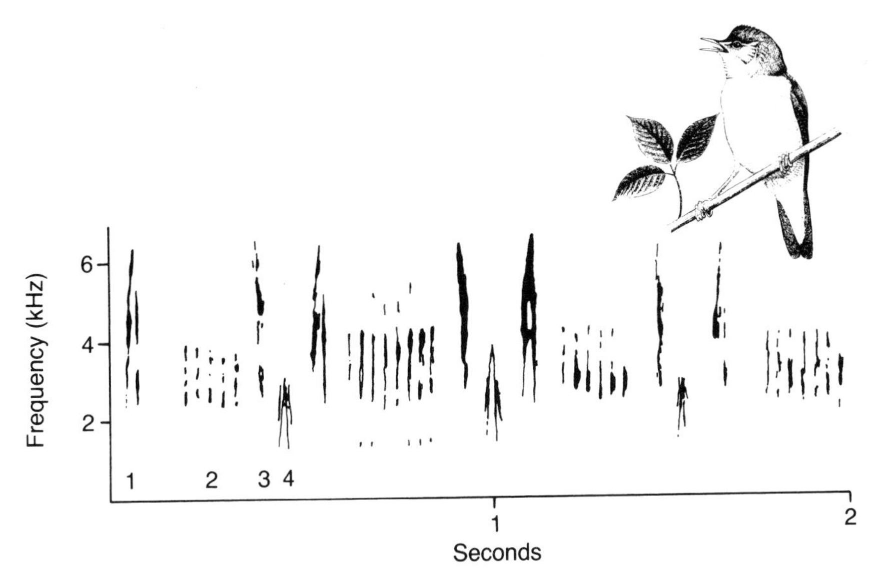

Figure 5.8. The song of an adult male marsh warbler, showing its complex construction. In this example, the elements that are repeated in the sequence shown are copied from four different African species: (1) tawny-flanked prinia, (2) African robin chat, (3) red bishop, (4) brown-headed tchagra. (From Dowsett-Lemaire, F., 1979. The imitative range of the song of the marsh warbler *Acrocephalus palustris*, with special reference to imitations of African birds. *Ibis* **121**, 453–68).

variations they show, in no case is it reasonable to think of development as entirely either genetically or environmentally determined.

5.5.2 Recognising predators and prey

All animals must distinguish between things that they can eat and things that they cannot; they must also recognise other animals that are likely to eat them. These tasks may be quite simple if the animal specialises on one or a few types of food, and if there are not many different predators that could eat it. On the other hand, species like rats and humans eat a wide variety of different foods and must distinguish those that are nutritious from those that are poisonous. Threats too can take many forms: a killer whale, a polar bear and a man with a gun look quite different but are all potentially lethal to a seal.

The extent to which food preferences are present at birth varies considerably between species. Newborn garter snakes show strong species differences

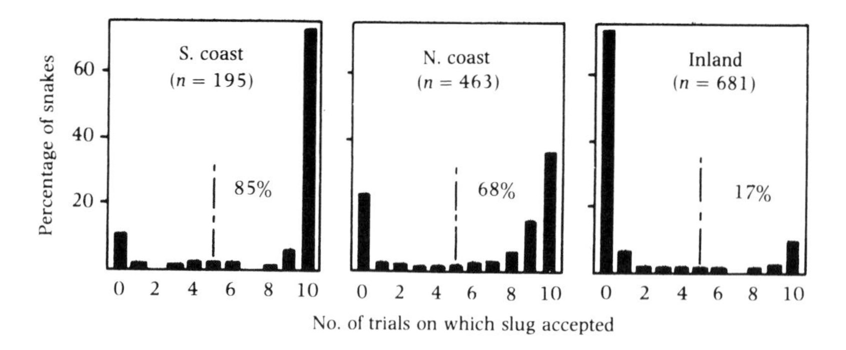

Figure 5.9. In California, the garter snake *Thamnophis elegans* feeds almost exclusively on slugs near the sea, but mainly on frogs and fish inland where slugs do not occur. Young snakes offered one piece of slug each day for l0 days show similar preferences, those from two coastal areas tending to accept them, while inland ones almost invariably reject them. These preferences do not result from experience as none of the snakes had encountered slugs before the tests began. (From Arnold, S.J., 1981. *Evolution* **35**, 510–15.)

in their responses to various food items which are in line with the preferences of adults. Even within a species, populations may differ. *Thamnophis elegans* populations near the sea in California specialise in eating slugs, whereas those inland eat frogs and fish. Newborn snakes can be tested for their preferences by seeing how many tongue flicks they make to extracts on cotton wool swabs. As Figure 5.9 shows, their preferences are in line with the area from which they come.

In other cases preferences can be acquired, and this enables animals to adjust their feeding to match availability, a skill at which rats are past masters.

Rats can learn about food for themselves or they may pick up information from others. If they encounter a new type of food, they will have a brief nibble and then wait for a period before returning to it. Should they feel ill during this time they will not come back but will reject the food thereafter. This is called the Garcia effect after John Garcia who discovered it. It is an excellent way of avoiding being poisoned, though it is infuriating to those humans who are bent on exterminating rats.

Rats seem to learn about food from others in a variety of ways. One route, curiously enough, is through their mother's milk. Young rats, when newly weaned, show a preference for the foods on which their mother had been feeding during lactation. Doubtless chemical cues about the nature of the food pass to them through the milk and help them to learn by her experience. Later in life, when living in their social groups, rats are also known to prefer foods on which one of their companions has been feeding. Odours clinging to the fur or detected in the faeces of the one that has found the food seem to be involved.

Eating food that is not very nutritious or makes one a bit sick is not a disaster, whereas a brush with a predator may be one's last. Again, evolution has led to various different developmental schemes which avoid this eventuality. In birds, some species appear to respond to specific features of predators such as hawks and owls without prior experience of them, while others take evasive action from all large objects that loom up to them and only slowly start to ignore those that prove to be harmless. Some neat experiments in this area have been those by Vilmos Csányi, a Hungarian ethologist, on paradise fish. Young fish of this species will swim towards and investigate either a pike or a goldfish introduced into their tank (Figure 5.10): a thoroughly dangerous enterprise in the case of the pike as it is likely to chase and catch them. If they are chased by it, they will avoid approaching on later occasions while still being prepared to swim close to the goldfish. However, a mild electric shock the first time they meet a goldfish makes them treat it with caution too. Thus these experiments illustrate a generalised response to various different objects at the outset which does not place the animal at too great a risk but predisposes it to learn which are dangerous and which are not. No matter what mix of predators there are in its environment, its experience enables the young animal to recognise them and take evasive action while continuing to feed peacefully when harmless animals pass by.

Learning for oneself about predators is a somewhat hit and miss procedure, given that failure to get it right the first time may be lethal. Learning by the experience of others is one way in which it can be improved. A good example here is how young rhesus monkeys learn about snakes, as described in Box 1.1.

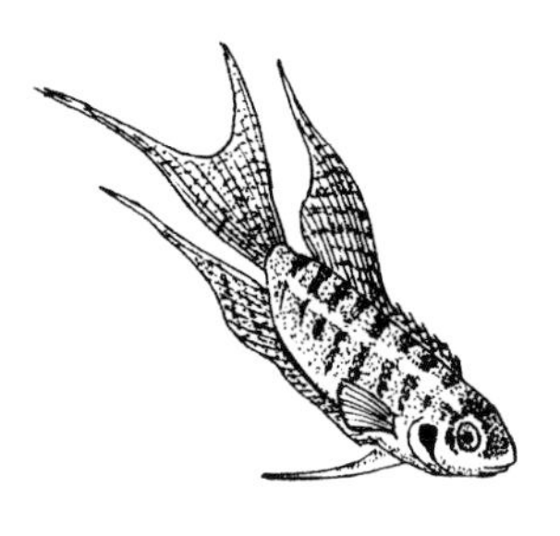

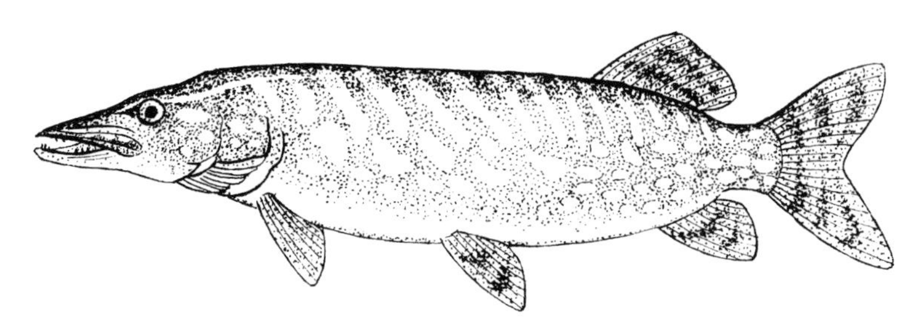

Figure 5.10. Paradise fish learn to avoid pike as a result of being chased by them. If exposure to a harmless goldfish is accompanied by a mild electric shock they will also learn to avoid this species.

Another case in point is the way in which young birds learn about predators, which was studied by Eberhard Curio and his colleagues. Birds of Prey are often mobbed by smaller birds, which chase them in flight or fly around them producing alarm calls when they are perched. The reasons for this behaviour have never been very clear, but it transpires that it passes from adults to young culturally and may therefore serve to indicate that a predator is a threat to younger individuals. Curio housed a young blackbird and an adult in adjacent cages, with a set-up between them which could be rotated so that the adult saw a stuffed owl, which it mobbed, while the young one saw a detergent bottle. Not only did the young one start to mob detergent bottles, but the habit was subsequently passed on to others in turn. Next time you are doing the washing up, if there is an irate blackbird at the window you will know where it came from!

5.5.3 Social development

Many animals live in groups and develop social relationships with other individuals of their species. Initially this may just be to the mother, to whom an attachment grows early in life, ensuring that the young animal stays close by and does not wander into danger. In birds this process is called imprinting. The young chick does not immediately recognise its mother, but will follow, learn about and imprint upon any large moving object that it is exposed to, be this a model chicken, a toy car or, in the most famous example, Konrad Lorenz. As with the learning of bird song, there is a sensitive period for imprinting: in birds such as domestic chicks and ducklings, which leave the nest shortly after hatching, this is the first few days of life. The chick may be more fussy about sounds, and there is some evidence that chickens prefer objects that cluck and ducklings those that quack, but generally its visual preference is, at least initially, for the largest and most striking thing around. Flashing lights are especially good. By following whatever object it has encountered for a few days it learns all about it and then seeks it when distressed and rejects other objects even when these look similar. If imprinted on a red watering can, it comes to prefer this to a blue one or to a red jug. In nature, of course, the result of this process is that it learns what its mother looks like and becomes attached to her; red watering cans do not normally wander round the territories of most birds.

This is, however, not all there is to the story. There is evidence that natural stimuli, such as a stuffed hen, though not particularly striking, may be more durable in their effects. Thus the preference for them is less easily overwritten by experience with attractive objects encountered later. Gabriel Horn and his colleagues found that chicks kept in diffuse light until past the time of imprinting and then exposed to a model chicken and a red box, as one might expect, show no preference between these objects. However, if they are first exposed to them after 24 hours in normal cages instead, they show a strong preference for the hen. They may not have had a preference at first, but they seem to have a growing one that is unrelated to experience. It turns out that the head and neck region of the hen are particularly attractive stimuli. If the hen is 'scrambled', as in Figure 5.11, its attractiveness to a chick depends on whether the head and neck are visible. Indeed, pitting a stuffed hen against a duck or a polecat, all of them with their heads and necks, yields no preference for the hen.

To Lorenz, a central characteristic of imprinting was that it taught the young animal the features to seek out in a prospective mate. Thus a chick imprinted on a red watering can is quite likely, when it grows up, to court and attempt to mate with one. More recent evidence suggests that this sexual imprinting does indeed take place, but at a later stage than filial imprinting, at a time when young birds are approaching maturity and can learn the features of a wide variety of animals, including their siblings, and not just those of the

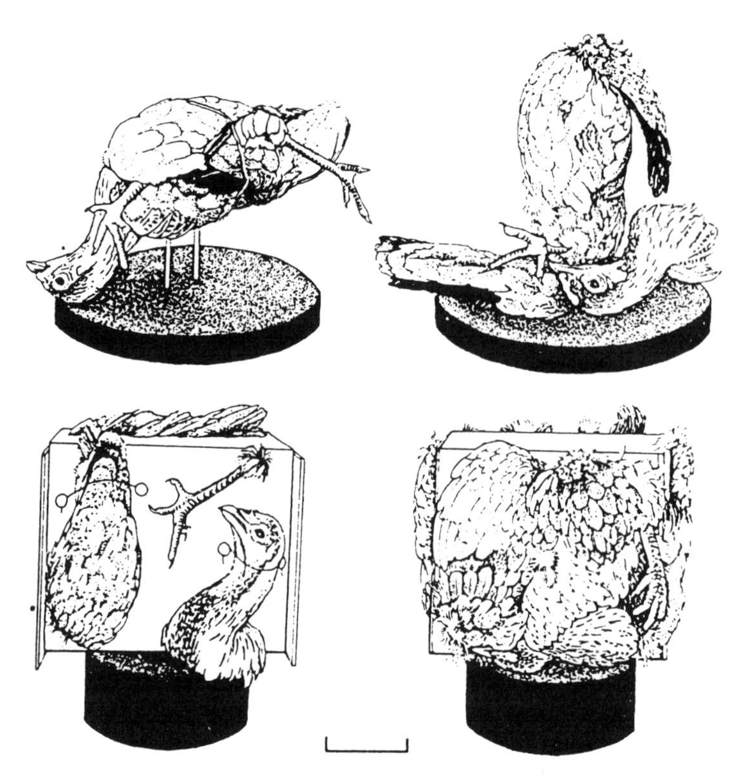

Figure 5.11. Some of the test objects used in imprinting experiments to assess the importance of hen-like features. Jumbled up bodies were as effective as a normal stuffed fowl provided they included the head and neck. The only one of these four models that did not was that on the bottom right, and it was the only one to which the chicks clearly preferred the stuffed hen. Scale bar = 10 (Drawing by Priscilla Barrett, from Johnson, M.H. & Morton, J., 1991. *Biology and Cognitive Development: The Case of Face Recognition*. Blackwell, Oxford.)

female that reared them. Given the differences in plumage between many male and female birds such as ducks, it would certainly not help young females to identify a suitable mate if they sought out birds that looked like their mother!

In mammals there is a process analogous to imprinting which leads to young ones being attached to their mothers. Sigmund Freud thought that this occurred through 'cupboard love', the mother rewarding the young one with milk so that it benefited from staying close to her. This is not so, however. In a classic series of experiments, Harry Harlow studied young rhesus monkeys reared with two mother substitutes (Figure 5.12). One of these was a frame covered with towelling to make it cuddly, the other was the frame alone but

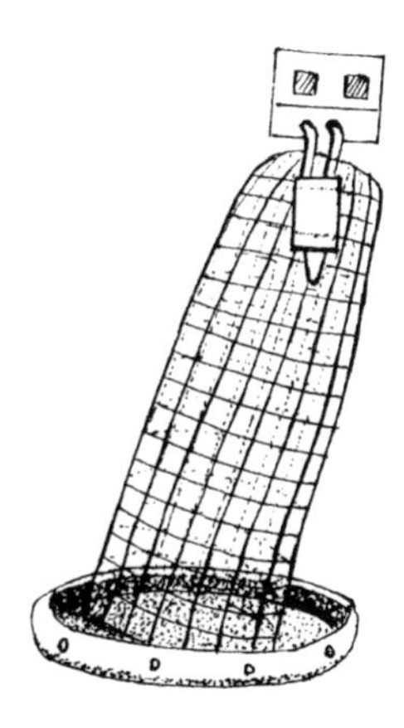

Figure 5.12. Young Rhesus monkeys isolated from other animals but reared with both a cuddly cloth-covered model mother and a wire one containing a milk bottle which feeds them become attached to the former, showing that attachment is not just a matter of 'cupboard love'.

with a bottle of milk mounted in it. When frightened the young animals rushed to the towelling mother and ignored that which provided milk. Indeed if the two were close together they learnt to cling to the cuddly one and lean across to drink from the other.

Rhesus monkeys therefore become attached to their mothers without the need to be fed by them, and they use them as bases from which to explore. The parallels between this growth of attachment and the imprinting of young birds are very interesting. Human babies, even a few hours after birth, have a predisposition to turn their heads towards pictures of faces rather than one with the same features but all jumbled up (Figure 5.13), and this no doubt helps them to focus on and learn about those around them. When they first start to smile at a few weeks old, they are rather unselective, and will do so to any face that appears. Later the face must be that of someone whom they know well, and often only that of their mother will do. Any stranger who has lifted a happy smiling 9 month old from its mother's arms is likely to know the dramatic effect this can have! Thus babies, like birds, start off being responsive to a wide range of stimuli, but with predispositions that focus their attention on appropriate ones. But later they become attached to those that they have experienced a lot and frightened by all others.

Baby monkeys spend a lot of time clinging to their mothers and being suckled, but as they become more mobile they begin to move about, exploring their world, meeting different adults and playing with their age-mates. The

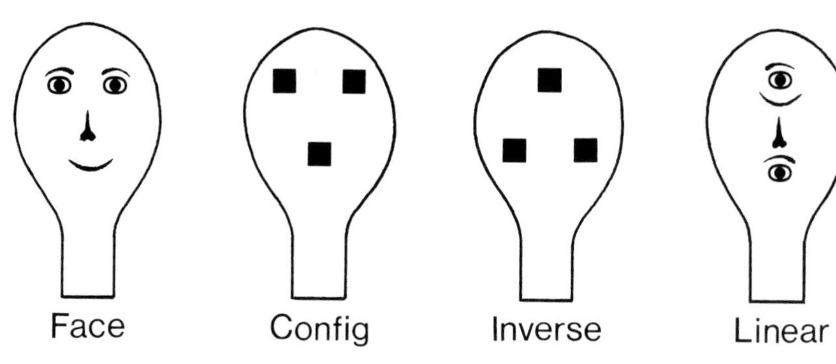

Figure 5.13. Four stimuli used to examine the tracking movement of newborn human infants. That labelled Face elicited more head and eye turning that those labelled Inverse and Linear, while the response to Config varied a lot from baby to baby. The results suggest that there is something attractive about the human face even to inexperienced infants less than 90 minutes old. (From Johnson, M. H. & Morton, J., 1991. *Biology and Cognitive Development: The Case of Face Recognition*. Blackwell, Oxford).

Figure 5.14. A baby rhesus monkey uses its mother as a base from which to explore the world.

mother is close by at first, keeping an anxious eye and grasping the infant if danger looms (Figure 5.14). But later the infant wanders further afield, only occasionally returning to her to feed and when in need of comfort. Interestingly, now their roles reverse: instead of the mother grasping the infant and restraining its wanderings, she tends to push it away and reject its approaches, as if encouraging its independence. This is a well-known phenomenon called 'weaning conflict'. The infant wants to go on feeding at the breast; the mother wishes to stop it doing so (and who could blame her as the young one's teeth erupt one by one!). Detailed studies by Robert Hinde and his colleagues on many different rhesus monkey mother–infant pairs have shown just how much their relationships differ from one another, and what an impact these differences may have on the later behaviour of the infant. Some mothers are possessive, some rejecting, some infants demanding, some independent, and these characteristics affect the quality of the relationship that they have with each other. Yet again it is clear that early experience can have a profound effect that lasts throughout life.

5.6 Social learning

A young male chaffinch copies his song from an adult singing nearby; a rat comes to prefer a particular food as a result of smelling it on the breath of a cagemate; a monkey becomes fearful of snakes after seeing the alarm of a companion confronted with one. All of these are examples of social learning in the broad sense of influences that social companions have on the learning of individuals. Within a social group such cultural influences are of enormous potential importance for, once a particular skill or piece of information is discovered, there is no reason why it should not spread through the group and persist long after its discoverer is dead.

Social influences on learning may have many different forms. At one extreme are processes, such as local enhancement, where the companion simply draws the attention of the animal to the stimulus which learns about it for itself. The animal does not learn from the companion, but learns more quickly as a result of the companion being there. At the other extreme is imitation, where the full motor pattern is copied by one animal from another. Bird song learning is clearly an example of the latter, although perhaps rather a special case as the ability is focused on this one particular task. Other examples are not common, and can often be attributed to simpler processes. It is certainly important not to jump to conclusions!

The way in which blue tits and great tits in the British Isles learnt to open milk bottles is a good case in point here (Figure 5.15). These birds are well equipped for the task, as they are attracted to bright objects and they gain access to sources of food such as nuts by hammering at them with their beaks. All it needed was for one of them to peck at a bright and shiny milk bottle top and, hey presto, there was a splendid drink of cream. But the remarkable feature of this innovation was that it spread rapidly through the population so that, within a few years, no milk bottle in Britain was safe. Mapping the spread of the habit showed that it did not arise by chance but moved out from certain focuses. It is clear that some birds developed the habit for themselves and

Figure 5.15. A classic case of cultural transmission, though one less well-established than it once appeared: the opening of milk bottles by titmice. Here a great tit is shown removing the shiny top from a bottle before drinking the cream (photograph © R. Thompson).

others in the area followed suit, so that it soon became widespread. But experiments suggest that the spread may well have resulted, not from watching others and copying their behaviour, but from discovering bottles others had opened and thereby learning that they were a source of cream. Having had that experience, and with hammering a behaviour pattern already in its repertoire, it would be but a simple task for the bird to gain access to the next sealed bottle it came across.

A neat way of examining social learning, called the transmission chain, has been employed by Bennett Galef. With Craig Allen he trained rats in groups of four to have a distaste for one of two sorts of food by making them mildly ill after eating it. They were then given a free choice of the two foods, and one rat was replaced by an untrained one each day until, after four days, the group consisted entirely of new animals. Replacements then continued for a further period, the rat that had been longest in the group always being the one to be removed. Figure 5.16 shows the results of an experiment that was carried on for eight days, on the last four of which none of the rats had received the original training. There was still a marked difference between the two groups. Local enhancement probably plays a part here, the rats tending to join others that are feeding at a particular food dish, but it is not essential. The transmission even takes place when the rats are fed separately from each other. As found in earlier studies, the smell of a particular food on another rat is sufficient to bias an animal towards that diet.

Social learning in this broad sense is a widespread and important phenomenon among animals. On the other hand, imitation, in the strict sense of learning to perform a particular act as a result of observing it done by another individual, seems to be rather rare among animals, though this is currently a very controversial area. Devising the ultimate experiment that shows it clearly and is not open to other simpler interpretations has become something of a parlour game. Surprisingly, the evidence so far seems to be better for birds and rodents than it is for primates, excluding humans, of course, which most would agree are rather good at it!

5.7 Conclusion

The development of behaviour is clearly a complex process and many different strategies have been described. Some of these are more flexible than others, so certain behaviour patterns develop very similarly in a wide range of environments, while others are much more affected by environmental differences. Ethologists have always been especially interested in animal

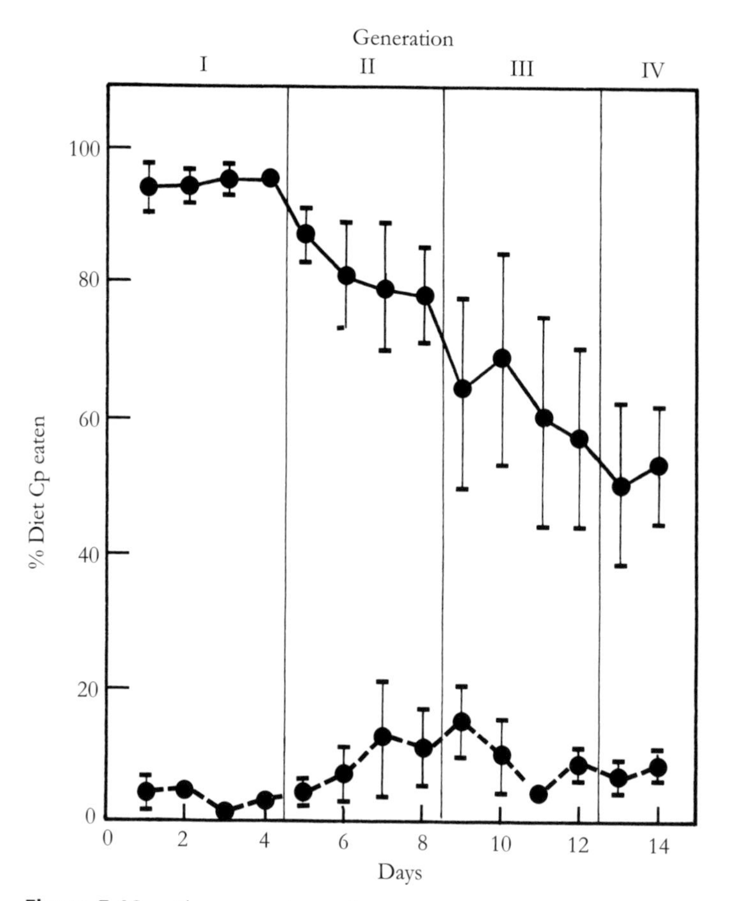

Figure 5.16. The percentage of cayenne pepper in the diet of colonies of rats fed both it and Japanese horseradish. The upper trace is the mean of five colonies trained to eat pepper, and the lower is that of five trained to eat horseradish. Each colony consisted of four animals, but every day one of these was removed and replaced by a naive rat. Thus the colony consisted of a mixture of original animals and replacements on days 2–4 and entirely of replacements, none of which had received the original training, thereafter. The continuing preference for the trained food is especially striking in the horseradish colonies. (From Galef, B.G. & Allen, C., 1995. *Anim. Behav.* **50**, 705–17.)

signals, such as courtship and aggressive displays. If these are to be understood by others, animals must get them right and thus they tend to be very similar throughout a species: hence the belief that they were fixed and 'innate'. Psychologists, with their very different interests, saw little fixity in what they studied. But now, with so many constraints on learning demonstrated on the one side, and with a great many environmental influences, including learning, shown to affect even the most constant of behaviour patterns, there is no longer this stark contrast of viewpoint. Genes and environment affect all behaviour and, as we have seen from many of the case histories in this chapter, discovering the way they interact with each other to produce an adaptive outcome has been one of the most interesting fields of animal behaviour in recent years.

6

Evolution

Behaviour leaves no fossils, though sometimes it is possible to surmise how animals in the past must have gone about their business. An animal with wings is likely to have been able to fly, and fossilised soldier termites will undoubtedly have defended their nests as do their present day equivalents. But such clues are rare and generally we must rely on studying how animals today behave to give us an idea about the evolutionary history of their behaviour. Two main approaches have proved particularly useful in this. First, is the study of the genetics of behaviour and, second, is the comparative method, looking at the behaviour of species alive today to reconstruct the changes that must have occurred during the course of their evolution. This last approach in particular has improved immensely in the past 20 years, so that our understanding of behavioural evolution has received quite a boost. It is no longer the poor sister among the four questions that it used to be. There are two main reasons for this. First, the use of molecular techniques has greatly improved our knowledge of the relationships between species, enabling behaviour to be fitted in to better established trees. Second, much better mathematical techniques for comparing across a range of species have been developed. Making such comparisons is not an easy matter, and only a flavour of what is involved will be given here.

We will start by looking at the genetic and comparative approaches in turn, and then go on to discuss the study of how displays evolved, a more speculative topic, but one on which a lot of ideas have been generated.

6.1 Behaviour genetics

In the last chapter it was argued that all behaviour is dependent on heredity, but the fact that genes affect a behaviour pattern is not sufficient for evolution to take place. Natural selection acts through *differences* in the success of individuals: those doing better leave more of their genes to the next generation than those doing worse so that good genes spread and the less good become scarcer. But, as it is the genes that pass from one generation to the next, this can only work if the differences between individuals are based on differences in their genes. Otherwise there is no way that the better behaviour can be inherited by the offspring of its possessor. To be strict therefore evolution requires not just that behaviour has a genetic basis, which all behaviour does, but that variations in behaviour are based in part on genetic variations on which selection can act.

Studies of the effects of genes on behaviour have taken several forms. Two of these will be discussed here: the examination of mutants and the use of selection experiments.

6.1.1 Studies of mutants

By selective breeding it is possible to generate a stock which is identical with that from which it was derived except for a single gene. Such is the *yellow* mutant in the fruitfly *Drosophila melanogaster*, so called because of the body colour change that it induces. Margaret Bastock, in a famous study, showed that males with this gene were slower at achieving mating than normal males and, on examining their courtship, she found that they displayed less of a component called vibration (Figure 6.1) which serves to stimulate the female. This single gene does therefore influence behaviour, in this case by altering its frequency. Such an effect may not appear very substantial but, of course, if the frequency of a behaviour pattern is modified down to zero it will not appear at all. Another *Drosophila* example concerns mutants at a locus known as *dunce*, because mutations there cause learning defects. In this case it is known that the product of this particular gene is an enzyme which breaks down the substance cyclic AMP and is particularly active in areas of the brain where learning takes place. This in turn suggests that the breakdown of cyclic AMP may be an important step in the storage and processing of learnt information, and one that is missing in *dunce* flies.

These examples may suggest simple connections between genes and behaviour, but numerous studies have shown that the links are very complex,

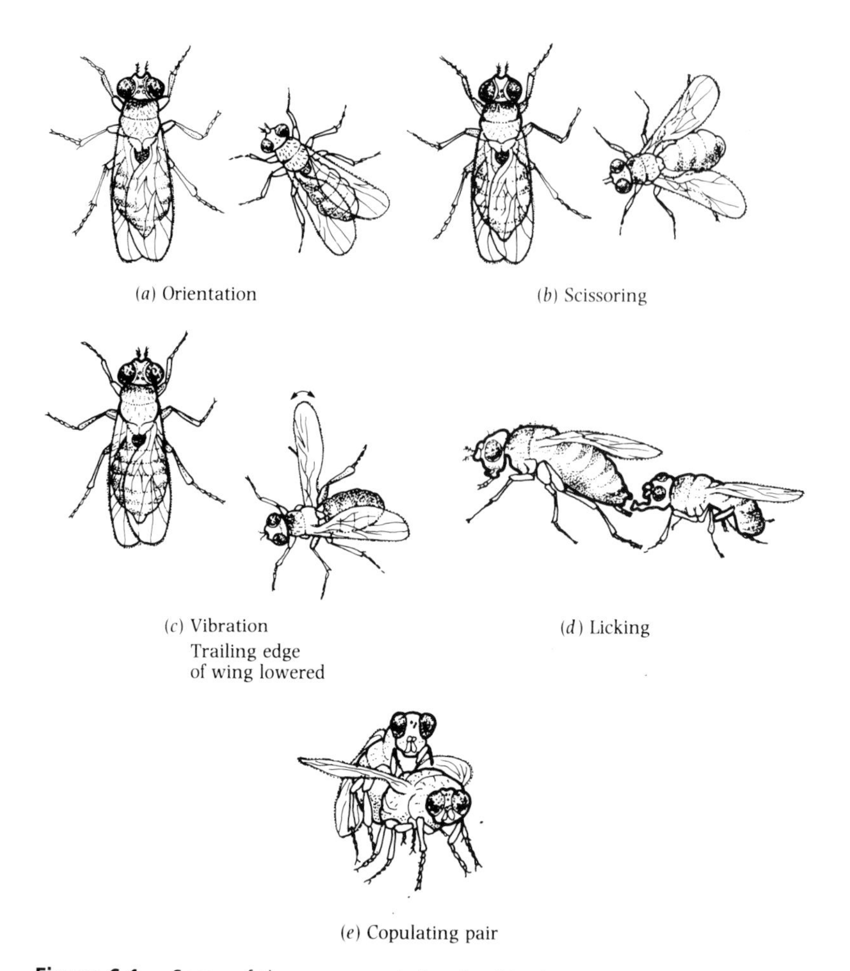

Figure 6.1. Some of the movements involved in the courtship of *Drosophila melanogaster:* (*a*) orientation of the male towards the female; (*b*) scissoring, in which he moves both his wings in and out to the side; (*c*) vibration, which involves the wing nearest the female being moved out and vibrated to and fro; (*d*) licking of the female's genital area by the male and, finally, (*e*) copulation. (From Manning, A., 1965. *Viewpoints in Biology* **4**,125–69.)

with each behaviour pattern influenced by many genes. A good case in point here is the way in which single gene mutations affect the activity of mice in a running wheel. Alberto Oliverio examined the effects on this of 31 different mutations, chosen because their presence altered coat colour and so was easily spotted, and showed that 12 of them altered the activity of their bearers. As there is no reason to think that coat colour and activity are likely to be linked to one another, this result suggests that about 40% of all genes probably influence activity in one way or another. This is certainly a far cry from the idea that one gene produces one behaviour pattern but, if one thinks about it, it is only to be expected. A feature such as activity is likely to be affected by innumerable factors: leg length, visual acuity, lung volume and muscular strength are a few that come easily to mind. These in turn depend on many physiological and biochemical processes and hence on a lot of genes.

Just as one behaviour pattern is influenced by many genes, so the opposite is also the case: one gene tends to influence many patterns of behaviour (a phenomenon known as pleiotropy). Again this should not be too surprising. Imagine a human gene which alters the level in the bloodstream of the male sex hormone testosterone so that it is much lower than in males lacking this gene. The mutant males are likely to be physically different, with less beard growth and more poorly developed muscles. The latter feature alone will probably affect many aspects of their behaviour, but some will be affected more directly, for testosterone itself is known to stimulate areas of the brain responsible for sexual and aggressive behaviour as well as the vocal apparatus to give deepening of the voice. The mutation would thus have widespread and substantial effects on behaviour.

All in all, then, the study of mutants has shown just how complex the interplay is between genes and behaviour. Many genes affect each behaviour pattern and a given gene is also likely to influence many actions. These points alone suggest that there will be plenty of genetic variability on which natural selection can act. Studies using artificial selection confirm this point.

6.1.2 Selection experiments

If a behaviour pattern can be altered by artificial selection, this indicates that variations in it must have a genetic basis on which selection can act. Many such experiments have been carried out and they have shown, first, that selection can change a wide variety of behaviour patterns and, second, just how these changes are brought about.

Once again, the fruitfly *Drosophila* has been specially useful here because a large number of generations of selective breeding can be achieved in a short period of time. Aubrey Manning selected for high and low mating speed in fruitflies. Out of 50 pairs he set up a line from the 10 that mated fastest and another from the 10 that mated slowest. In the fast line he always bred only from the 20% that mated most quickly in each generation and in the slow line he selected the 20% that mated slowest, carrying on for many generations.

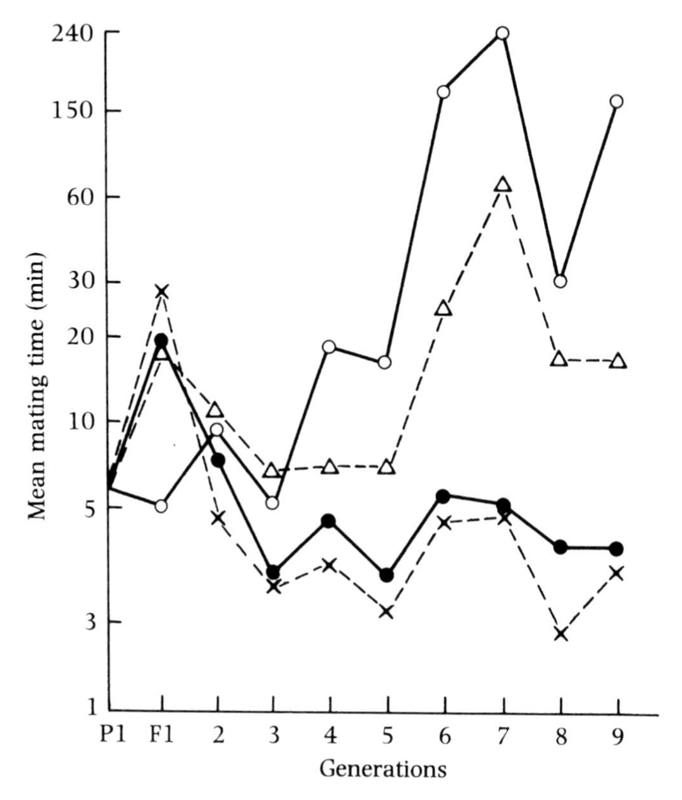

Figure 6.2. The results of the first nine generations of selection for fast and slow mating speed in *Drosophila melanogaster*. The points shown are the mean mating times for two fast lines and two slow ones. Note that the time scale is logarithmic, so that the fast line ended up mating on average about ten times quicker than the slow. This difference persisted without any major further change in subsequent generations. (After Manning, A., 1961. *Anim. Behav.* **9**, 82–92.)

The most dramatic effect was reached after just seven generations, with the fast line averaging 3 minutes to mate while the slow line took 80 minutes (Figure 6.2). Thus selection had had a dramatic effect on mating speed. But how had this been achieved? Examination of the behaviour of the flies showed that several changes were involved. The main effect in the fast flies had been to make them much less active, the males settling down quickly to vigorous courtship, while the females also showed greater responsiveness. In the slow line, by contrast, both sexes showed prolonged activity after being placed together and they often rushed past each other without attempting to start any interaction.

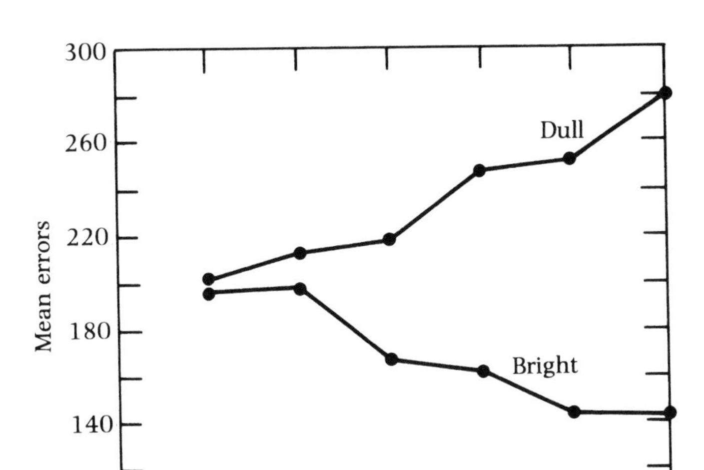

Figure 6.3. Results of an experiment in which rats were bred for maze brightness and dullness. After six generations, the dull rats were averaging around twice as many errors before learning the maze than were those in the bright line. (From Thompson, W.R., 1954. *Proc. Ass. Res. nerv. ment. Dis.* **33**, 209–231.)

Another good example comes from the experiments of Theo Bakker on aggression in sticklebacks. He successfully selected for lines of fish that were highly aggressive as adult males, as adult females or as juveniles. He found the last two to be linked, so selecting for female aggression also led to more in juveniles, while male aggression could be enhanced without changes in that of the other groups. The likely reason here is that male aggression is under the influence of male sex hormones, and selecting for more aggression in males enhances the level of these. In line with this idea, Bakker found that his selection for male aggression had also altered kidney size, another characteristic that depends on male sex hormones.

It is hard to conceive of a behaviour pattern in which there is no genetic variation on which selection could act. A good example here, which produced results which were surprising at the time, was experiments involving selection for learning ability in rats. Rats can be taught to run mazes for a reward of food at the far end, and some of them learn more quickly than others. Breeding from those that learn best and those that learn worst leads to a progressive divergence over the course of several generations (Figure 6.3). This shows that differences in learning ability among rats are to some extent genetically based, otherwise selection could not accentuate them from generation to generation.

These selection experiments show that those studying behaviour genetics are able to select for virtually any characteristic they care to choose. The ease with which they can do so depends on the extent to which variation in behaviour has a genetic basis; the way selection acts is to alter those features for which genetic variation is greatest (in jargon terms, the features that have highest heritability). But the fact that selection experiments on behaviour are successful should not be at all unexpected, given the multiplicity of genes that affect almost all behaviour patterns directly or indirectly. However, picking up a point from the last chapter, one should not confuse this with genetic determinism. Great play has been made recently about the genes that have been found to influence intelligence (or, indeed, homosexuality) in humans. It would be much more surprising if such genes did not exist! But, then, we must not forget that the environment is just as important in influencing what behaviour patterns are expressed.

From the point of view of this chapter the important message is simple: there is ample evidence of variations in behaviour being affected by genetic variations on which natural selection could act.

6.2 Comparative evidence

Although we cannot go back in time to see how behaviour evolved, we can look at different species alive today and ask how their behaviour has diverged since they split off from one another. This is obviously most fruitful if the species involved are closely related so that the differences between them are quite slight and the changes that have taken place are reasonably easy to reconstruct. Even then it is not absolutely straightforward for several reasons hinging on the fact that behaviour is very flexible. Here we will just mention two.

One point is that behaviour can diverge for reasons totally unconnected with genetics. We have already discussed examples in the last chapter of how behaviour, such as food preferences in rats and song in birds, can be passed from one individual to another by cultural transmission. This can lead to a particular aspect of behaviour being found in one population but not in another without there necessarily being any genetic differences between the two. Cultural evolution can take place very rapidly. For example, 20 years is long enough for virtually all the songs sung by a population of chaffinches to have been replaced by new ones simply because of the turnover it involves.

Likewise, in a famous example, one group of Japanese monkeys developed the habit of washing sweet potatoes before they ate them because one individual in the troop discovered the trick and it then spread by social learning to most of the others. Marked and rapid changes such as these, which are often adaptive, can occur without genetic changes.

A second problem of interpreting behaviour differences between species is that of convergence. This is always a difficulty in interpreting evolutionary evidence, but it is particularly so with behaviour because of its flexibility. Two features may be similar because they are derived from a common ancestor, in which case they are referred to as homologous, or they may be so because their bearers have converged upon the same solution to a common problem, so that they are simply analogous. The wings of birds, the arms of humans and the pectoral fins of fish are homologous, all being examples of the vertebrate fore-limb. But the wings of birds and of insects, while serving similar functions, are only analogous with one another as they are not derived from common structures. In the case of behaviour, distantly related species may have similar characteristics because they rely on similar food or they live in much the same habitat. This may seem like a problem, but it is actually part of the interest of the subject. Especially nowadays, when we often have detailed information about how species are related to each other, we can often discriminate between convergence and common ancestry. Although some of them do look forward to the topic of function discussed in the next chapter, we will discuss the various comparative approaches to understanding behavioural evolution and adaptation here.

6.2.1 Comparing one species with others

Niko Tinbergen's student, Esther Cullen, did her doctorate perched on top of a cliff watching kittiwakes, a small gull that nests in dense colonies on cliff ledges in the higher latitudes of the northern hemisphere. This nesting place is very different from the open areas where most other gulls breed (Figure 6.4) and the kittiwake's behaviour also differs from theirs in many different ways. Cullen concluded that the differences could be traced back to consequences of the cliff nesting habit.

The number of ledges on cliffs is limited and, in keeping with this, kittiwakes guard their nest sites jealously, forming their pairs there well before egg-laying and doing all their courting and mating on the nest. Other gulls nest in open areas where sites are numerous and so they only need move to them just before laying as they are not in short supply. The kittiwake's nesting place is

Figure 6.4. A colony of kittiwakes at their nests on a cliff face. Note various ways in which this species differs from ground nesting gulls: the legs are short so that balance is easier, the ledge has been built up with mud thus making it safer, the chicks are conspicuously patterned and droppings are allowed to accumulate around the rim of the nest. These last two features are thought to have resulted from lack of predation pressure on birds nesting on cliffs (photograph © J.C. Coulson).

also much more dangerous than that of other gulls and, doubtless because of this, their nest building is more elaborate. They enlarge the ledge with mud and vegetation and scrape out a deep nest cup. Not surprisingly for animals living in such a precarious position, both adults and chicks have short legs and strong claws and, during copulation, the female sits rather than standing as do other species. The young ones flap their wings very little when fledging approaches and, even if attacked, they will not leave the nest. With a fall of hundreds of feet just beside them, few would doubt that this is functional! Another difference from other gulls probably also arises because the nests are on separate ledges and the young cannot move about between them: adult kittiwakes do not recognise their own chicks, presumably because they simply do not need to.

Esther Cullen attributed another set of kittiwake features to the fact that few predators can get at nests on cliffs so that, while they are dangerous in one way, they are safe in another. As one would predict from this, kittiwakes seldom produce an alarm call. The nests are also not camouflaged like those of other gulls: thus the droppings of the young are not removed and the chicks themselves are far from cryptic, being strikingly patterned in grey, black and white.

This analysis of the relationships between all these different features is a convincing one, although one cannot be certain about it as all Cullen could do is point to correlations between them and the fact that kittiwakes nest on cliffs. Her conclusions are made firmer, however, because some other gull species, not closely related to the kittiwake, also nest on cliffs and have the characteristics that go with it. The Galapagos swallow-tailed gull is an example here.

Another contrast between closely related species came to light from a study by Tim Clutton-Brock of two species of monkeys in East Africa (Figure 6.5). The black and white colobus and the red colobus overlap in their ranges and both live in forests where their diet consists primarily of foliage. In many other ways, however, they are very different. The red colobus is restricted to wet forests where it lives in troops of 40 or more, ranging over about a square kilometre. They feed on the flowers, shoots, fruit and leaves of many different types of trees. The black and white, by contrast, lives only in dry forests where it feeds on the mature leaves of a few tree species. Its troop size is usually fewer than 10 and its range much more restricted than that of the red.

What is likely to have led to these differences? Clutton-Brock argued that the primary factor may be that the red colobus requires a more varied diet. If this is so it can only live in wet forest as it needs an area where some trees are in fruit at all times of year. This requirement will also constrain it to feed on many species and to have a large range so that food will always be available.

Figure 6.5. (*a*) The red colobus (photograph © T.H. Clutton-Brock) and (*b*) black and white colobus (photograph by D.J. Chivers): two species of monkey that have been found to differ considerably in their group and home range sizes, probably as a result of their very different feeding requirements.

Having a bigger group enables efficient usage of resources in a range which is large and helps in its defence.

This is a plausible argument and again, as with the kittiwake, it suggests that substantial differences in the behaviour of related species may arise for rather slight reasons. It is not, however, so easy in this case to be sure about cause and effect. There is a lot of argument involved in the claim that the diet is the main reason for the differences between the species and some of the threads of this could run in quite different directions. For example, a larger home range may make it possible to eat fruit at all times of year rather than being a result of the need to do so. All that is known is that certain features are correlated, but it is not easy in cases such as this to carry out experiments to discover which causes which.

6.2.2 Broad cross-species comparisons

The way in which animals have evolved so that their behaviour matches them to their way of life can be examined by studying the distribution of different characteristics in a group of species. This approach involves looking at the correlation between different features to see if they are linked. As pointed out in Section 1.3, correlations can be interpreted in various different ways, and one has to be careful not to jump to conclusions. It is so much the better if one has clear alternatives with different predictions before the analysis is carried out.

A few primate examples will help to illustrate this approach. Among primates males are often larger than females, and various theories have been put forward as to why this should be the case. One idea was that it avoided feeding competition, as the smaller females were able to exploit resources, such as the leaves at the end of twigs, that the bulky males could not reach. Another was that males competed with each other to mate with females, and being larger enabled them to fight off rivals more successfully. If this last idea was true, we would expect the difference in size to be greatest in groups where there was more competition, for example where one male mates with several females and other males are excluded. On the other hand the feeding competition hypothesis would make no such prediction. As the plots in Figure 6.6 *a* and *b* show, dimorphism in body size, and in that of the canine teeth which are used in threat and in biting, is indeed greatest in species where one male monopolises a group of females. The breeding competition hypothesis is thus the more likely of the two to be correct, though differences in diet may also be important.

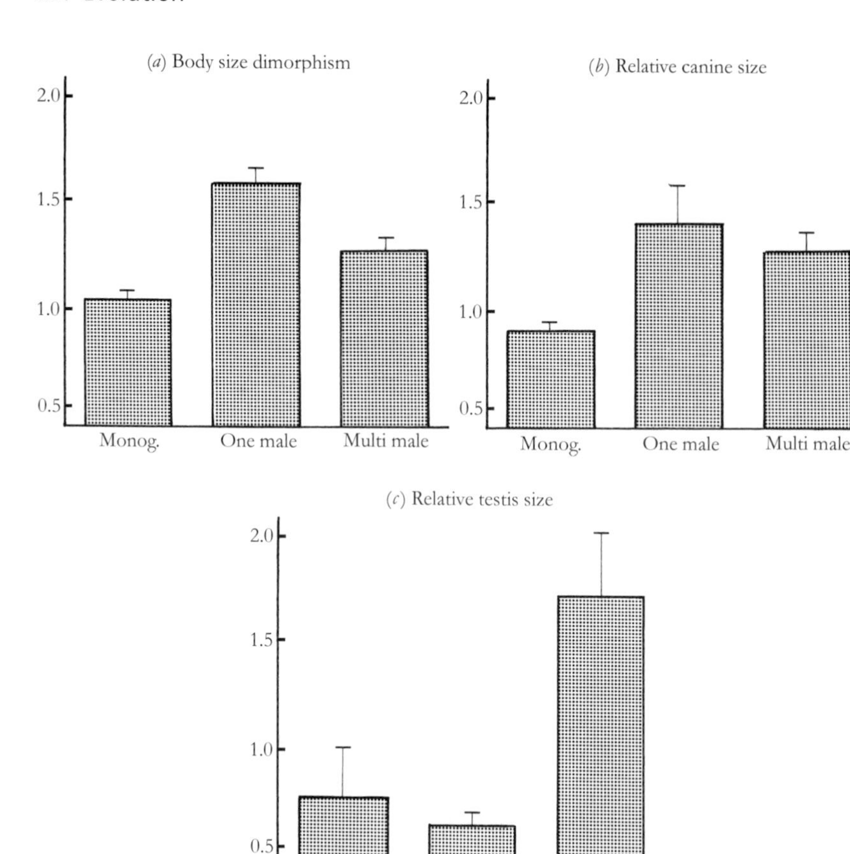

Figure 6.6. The relative sizes of bodies, teeth and testes in primate species split according to whether the species are monogamous, form groups in which one male monopolises several females, or occur in multimale groups in which there are many males and females. The measure of 'relative size' allows comparison between the breeding systems regardless of the species' absolute size. Each bar shows the mean (and standard error, an indication of variation) of several species. (After Clutton-Brock, T.H. & Harvey, P.H., 1984. In *Behavioural Ecology: An Evolutionary Approach*, ed. J.R. Krebs & N.B. Davies. Blackwell Scientific Publications, Oxford.)

Figure 6.6*c* illustrates the test of another hypothesis, again on primates. It has been suggested that sperm competition takes place where fertile females mate with more than one male. A monogamous male, or one which has a group of females to himself, need only produce enough sperm to ensure that all their eggs are fertilised. On the other hand, where two or more males mate with a female, the most likely one to be successful in fertilising her eggs is that producing the most sperm. On this hypothesis males in the more promiscuous multi-male groups should have larger testes than those where males do not compete for fertilisation. Again, the results in Figure 6.6 fit in with this idea.

A difficulty with analyses such as these is to decide on the taxonomic level at which they should be conducted. It might seem obvious that the species, being the most well defined and clear unit in evolutionary thinking, would be the best one to use, the behaviour of each species being taken as a separate example. But there is a difficulty here. Some groups have a lot of species in them and some rather few. If we look just at the apes, for example, there is one species of human, of gorilla and of orangutan, there are two of chimpanzees and there are nine of gibbons. If we look at relationships among these species, we will find that small size, swinging through trees, long term pair-bonds and singing duets in the early morning are strongly related. But this is simply because these features are almost entirely restricted to the gibbons, all of which show them. But they did not evolve nine separate times: in all probability they did so once and existed in the common ancestor of all the gibbon species alive today.

Problems such as this have best been overcome by the methods of phylogenetic reconstruction.

6.2.3 Phylogenetic reconstructions

If we know the relationships between different species, we can construct a family tree and then, looking at the behaviour of each species within it, attempt to reconstruct how it must have changed during evolution. An example from displays will be given in the next section, but here we will consider behaviour patterns that are less distinct between species. For example, imagine that we find two closely related species one of which is a herbivore and the other a carnivore. Clearly some change has taken place in one or both lines since their common ancestor, but what is that change most likely to have been? One approach, called 'out group' analysis, is to look at the closest relative of the two species that one can find. If that is a herbivore, it is most parsimonious to suggest that the common ancestor of the two we are interested in was also a herbivore (Figure 6.7*a*). On that hypothesis carnivory has evolved once in the group of three species whereas, had the common ancestor of all three been a carnivore, one would have to propose that two evolutionary changes towards herbivory had taken place (Figure 6.7*b*).

Working back through a phylogenetic tree in this way can allow quite

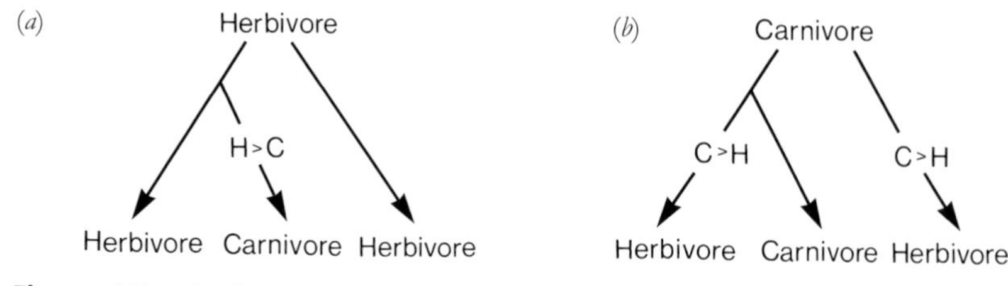

Figure 6.7. Analysing evolutionary changes by studying closely related species. Here two close relatives differ in diet, one being a herbivore and the other a carnivore, while their next closest relative is a herbivore. Assuming that the common ancestor of all three was a herbivore is most parsimonious as it requires only one evolutionary change (*a*) compared with two if it was a carnivore (*b*).

detailed reconstruction of how changes must have occurred. John Gittleman did this to look at the evolution of parental care in fish. There are four possible modes here: no care, paternal care, maternal care and care by both sexes. Examining the tree, Gittleman found 21 cases where a switch from one to another must have taken place during evolution (Figure 6.8). Interestingly, in all these cases the switch involved only the addition or subtraction of care by one partner: there were no switches from no care to both caring or vice versa, and there were none from one sex caring to the other sex caring. These changes were less likely, he argued, as they would have involved two steps rather than one.

These few examples are relatively simple. Many other comparative studies have now been carried out, and a variety of clever techniques have been devised which allow one to examine how behaviour has changed during evolution despite the fact that related species, like the gibbons, tend to behave rather similarly so cannot be viewed as separate.

6.3 The study of displays

Many of the most satisfactory examinations of the evolution of behaviour have involved the study of displays. As one of their prime functions is in mate recognition, they tend to be rather fixed and constant within a species and yet to differ between them. Indeed, in many closely related species observing courtship displays may be a much easier means of separating them than attempting to do so from morphology. Although the form of displays is often related to the habitat to ensure effective communication, as we shall see in Chapter 8, the differences between them are not so much affected by the

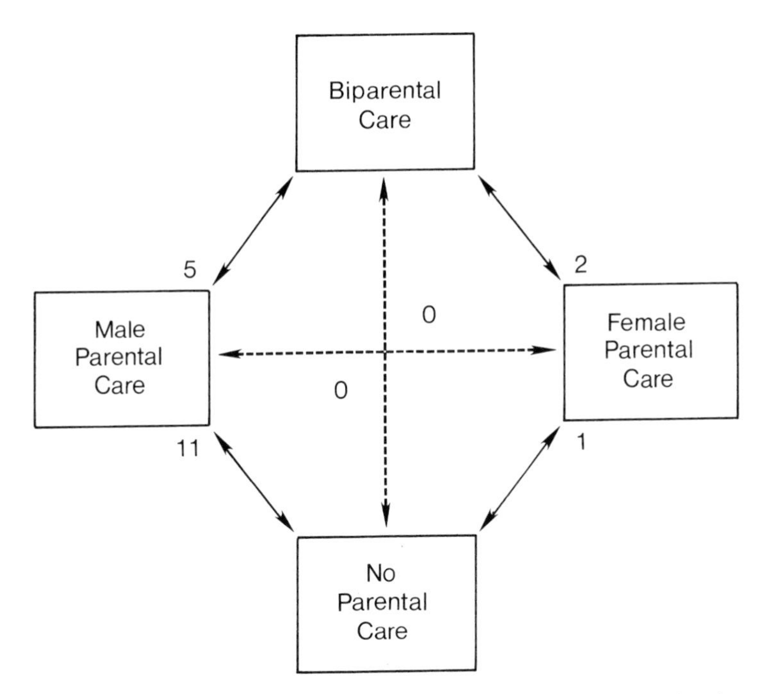

Figure 6.8. Four possible forms of parental care in fishes. Switches between them round the periphery of the diagram require a change in only one sex, whereas those across the centre need a switch of behaviour by both and were therefore suggested to be less likely. The figures give the number of genera or families in which the two forms of care were both present and in which a switch must therefore have taken place. None of these involved changes in both sexes. (Data from Gittleman, J.L., 1981. *Anim. Behav.* **29**, 936–41.)

environment as in many other actions, so convergence is less common and similarities between species are usually proportional to the closeness of their relationship. All these points make displays an easier subject for comparative study than many other aspects of behaviour.

Comparisons of displays have been especially useful to reconstruct the course of evolution at two different levels: the divergences that have occurred at the level of the family tree, and the more subtle changes that have taken place in the course of two species splitting off from one another. Having considered each of these we will then discuss where displays are thought to have come from in the first place.

6.3.1 Family trees

Konrad Lorenz pioneered studying behaviour as a means of comparing between species, arguing that it could be used in just the same way as structures to work out the relationships between them. He looked at various ducks and geese and found some behaviour patterns which they all had in common, like the monosyllabic piping call of the chick, and others which occurred only in a more restricted group, such as the 'burping' which appears among the displays of ducks but not those of geese. As displays are often elaborate, it is extremely unlikely that exactly the same one would arise twice during evolution, so one can assume that burping occurred in the common ancestor of all ducks and the piping call in that of the whole group. While displays may be lost as well as gained during evolution, the idea behind comparisons such as this is quite simple: two species that share a feature are more likely to be closely related than is either to a third species lacking that characteristic. If many aspects of behaviour are looked at, compelling evidence can accrue about the way in which evolution led to the differences between the species being studied.

A simpler case than that of the ducks and geese, is that of the displays of pelicans and their relatives (Figure 6.9), studied by a Dutch ethologist, G. F. van Tets. This is rather a diverse group and their displays differ quite a bit. Some are common to all species, some occur in only one and some in a small subgroup of the species. If one assumes that each display arose only once, then it must have been present in the common ancestor of all the species showing it today. This allows one to draw a family tree suggesting when the species split off from one another. As one would hope if the method was well founded, the tree that fits best is the same as that previously suggested on the basis of the structures of these species, confirming the idea that behaviour can be a useful guide to evolutionary sequences. With 37 behaviour patterns and 20 species in total, it is not surprising that there is not a perfect fit, but a recent detailed phylogenetic analysis by Martyn Kennedy and colleagues suggests that there are rather few instances where one or two species have lost a character or a feature appears to have evolved twice.

Sometimes animals may have curious behaviour patterns the origin of which it is hard to imagine but, by looking at related species, one can begin to see how they might have arisen. One must be careful, as no living species is the ancestor of any other, but some may have changed less than others since they separated so may give one an idea of what the ancestral form was like. A striking case here is that of the Empid flies. In this group different species show a wide diversity of related courtship displays. The most perplexing of

(*a*)

	Pelicans	Gannets & Boobies	Cormorants & Darters	Frigate birds
Kink throating (KT)			+	
Head wagging (HW)		+		
Sky pointing (SP) and Hop display (HD) and Wing waving (WW)		+	+	
Bowing (BOW)	+	+	+	
Presenting Nest material (PNM) and Prelanding call (PLC)	+	+	+	+

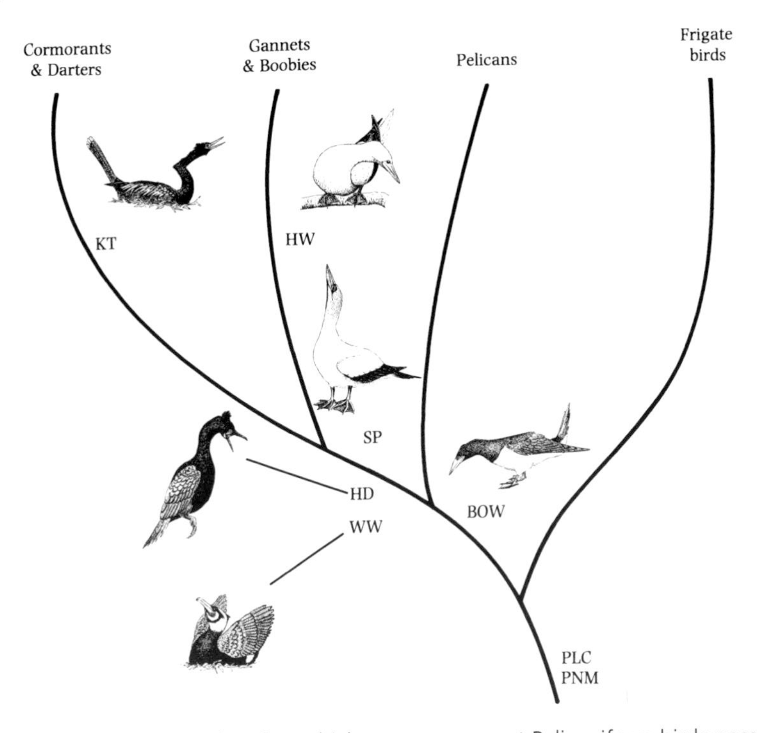

Figure 6.9. (*a*) Table showing which groups amongst Pelicaniform birds possess various different displays. (*b*) A tree showing the pattern of evolution of the group based upon morphological features with the point at which the displays are likely to have first appeared indicated on it. The distribution of the displays is compatible with the morphological evidence: it is not necessary to make the unlikely assumption that a single display evolved twice. (After van Tets, G.F., 1965. *Ornithological Monographs* **2**, 1-88.)

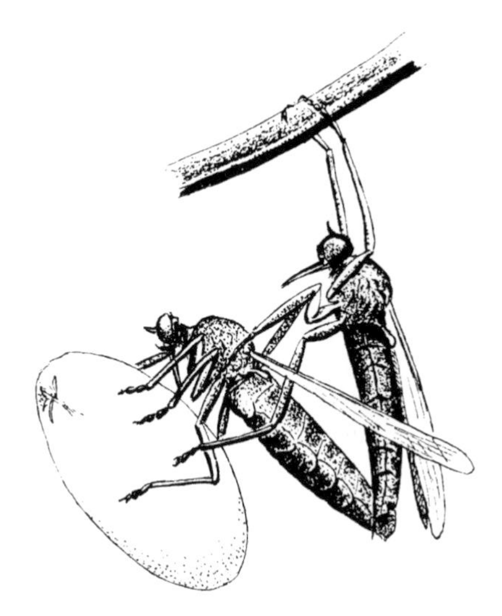

Figure 6.10. Before mating, the male of the empid fly *Hilara sartor* courts the female by presenting her with an empty silken balloon. Here a pair are shown mating, with the female on the left clutching the male's gift.

these is in a species where the male spins a delicate balloon of silk before joining the swarm in which mating takes place (Figure 6.10). In the swarm he presents his balloon to a female and, as if it symbolised a wedding contract, her acceptance of it is a prelude to mating. How on earth could such a bizarre ritual have evolved? E. L. Kessel argued that, if one looks at related species, one begins to see a sequence which could have led to this strange behaviour. There are seven species and they behave in the following ways.

1 The male simply courts the female without giving her any token.
2 The male captures a fly and presents it to the female.
3 The male captures a fly and wraps it in a little silk before giving it to the female.
4 The male captures a fly and parcels it totally in silk before giving it to the female.
5 The male captures a fly and sucks it dry before wrapping it in silk and presenting it to the female.
6 The male picks up a fragment of dead fly and bases a silk parcel upon it.
7 The male spins a balloon of silk without using any prey as a basis.

The first five species feed on insects and so those of them that present a gift may be conferring a real benefit on the female. The last two, however, are nectar feeders. Their gifts are of no intrinsic value but evolution has left females with a preference for males that present gifts so, within a swarm, only males that do so are likely to get a mate. The whole procedure has become a ritual. It is a good example of the action of sexual selection, to which we shall return in Section 8.5.

Courtship feeding is quite common among animals and may often help the female to lay more or better eggs so that it benefits both her and the male that mates with her. Though it is a scorpionfly rather than an Empid, there is a species in which males mimic females and approach other males which are bearing their gifts and looking for mates. By putting on all the appearance of being receptive, these mimics are presented with dead flies just as if they were females. Having received their gift they then fly off in search of a female of their own!

6.3.2 How displays diverge between species

Animal species can be defined as groups of individuals that actually or potentially interbreed with one another. Members of different species tend not to pair with each other in the wild and one of the main reasons for this is that they have distinctive displays which advertise the exact species to which they belong. It is usually highly disadvantageous for animals to mate across species for, even if hybrids are formed, these tend to be either sterile, like mules, or at a disadvantage compared with pure-bred animals. Thus it is in the interests of males to produce clear, species-specific signals and of females to reject those that are not exactly right.

There are several ways in which changes may occur in courtship displays so that species become distinct from one another. We can return to fruitflies to illustrate some of these. When these little flies court the male orients towards the female and displays to her with wing movements which usually consist of extending the wing nearest to her and vibrating it up and down (see Figure 6.1). This drives a stream of air towards her, which may well bear a stimulating odour, but it also makes a sound that can be heard if a pair of flies is induced to court on top of a sensitive microphone. During evolution, the exact form of this courtship has come to show differences between species in several different ways:

1 Changes in threshold. Two closely related species, *Drosophila simulans* and *D. melanogaster*, show this well. When pairs are put together the male *simulans* takes longer to begin courting and is more sluggish in his actions, but the female is more receptive so counteracting the effect. Her threshold for accepting the male is thus lower.

2 Change in the senses used. The courtship of *simulans* is slowed down in the dark, whereas that of *melanogaster* is not, suggesting that visual cues are important to the former but not the latter.
3 Changes in the frequency of an action. In *melanogaster* vibration occurs frequently, while a movement of both the wings together known as scissoring is very rarely shown. By contrast, *simulans* shows more scissoring and less vibration.
4 Changes in sequence. Flies of the group to which *melanogaster* and *simulans* belong show these wing movements before mounting, while in the group to which *Drosophila rufa* belongs vibration occurs once mounting has taken place.
5 Timing changes. These are best illustrated by the sounds produced by different species. Both the interval between pulses and the frequency of the sound within a pulse differ between species. Interestingly, in cases where one of these features is the same in two species the other is very different and this is probably sufficient to isolate them from each other (Figure 6.11).

Each of these differences is slight in itself but a combination of them, or a succession of changes in one of them, can lead to marked contrasts between species. As was mentioned earlier, a display which becomes progressively rarer may eventually cease to appear at all, so the presence or absence of behaviour patterns may depend on such gradual shifts.

6.3.3 The origin of displays

Comparison between closely related species can therefore indicate how gradual changes have taken place during evolution. A more difficult question is where these displays came from in the first place. Presumably animals did not suddenly start leaping up and down and waving their arms around for no apparent reason, yet actions such as these may be crucial to their courtship today. Several ideas have been proposed here, based largely on the comparison of displays with the everyday behaviour of the species showing them. If an action looks similar in the two contexts then perhaps the one was derived from the other and we can ask why this particular act rather than any other should have become incorporated into the display. Of course many displays may be very ancient so that comparisons with current behaviour may be somewhat tenuous: the wing-flapping display of a bird may be derived from the movements made by some ancestral fish with its pectoral fins! Nevertheless, some quite convincing links have been found between displays and other actions (Figure 6.12).

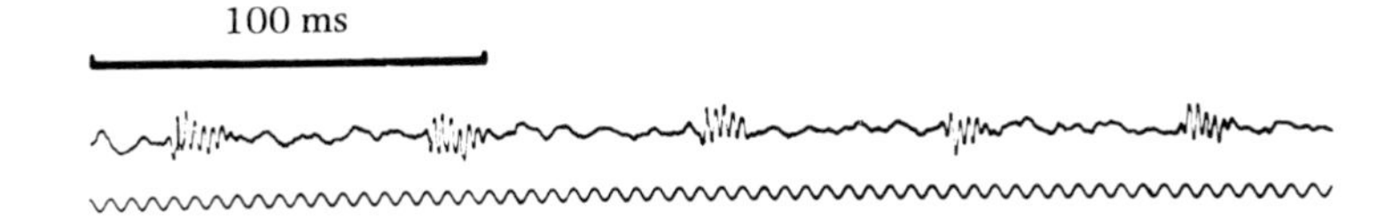

(*b*)

Species of *Drosophila*	Frequency of pulse (cycles per second)	Interval between pulses (milliseconds)
pseudoobscura	258 ± 11	37.9 ± 6.8
persimilis	528 ± 16	53.7 ± 4.3
ambigua	294 ± 35	38.4 ± 3.7
algonquin	226 ± 24	53.3 ± 10.9
athabasca	444 ± 15	11.7 ± 1.1

Figure 6.11. (*a*) Oscilloscope trace of the sound produced by the wing vibrations of a courting male *Drosophila persimilis*. (*b*) Table to show how two features of these sounds vary amongst five closely related species from the group to which persimilis belongs. All are distinct from each other in one or other of these measures with the possible exception of *pseudoobscura* and *ambigua*. That these two are so similar is of no consequence to selection as the former is European and the latter North American. (From Ewing, A.W., & Bennet-Clark, H.C., 1968. *Behaviour* **31**, 288–301.)

On the basis of such comparisons, the following general categories of action are thought to have given rise to displays.

1. *Intention movements.* Before they perform some action animals often show movements which indicate that they are getting ready to do so. For example, a bird about to take off crouches low on the ground with its muscles braced and its wings out from the body, all set to propel itself up and away. Intention movements are particularly common as a preparation for locomotion, and displays often bear a strong resemblance to them. One reason for this is probably that such actions are common when two animals are close to each other, as they are during courtship. For most of the year, animals tend to keep their distance, avoiding getting near enough to others to risk being pecked or bitten. But you cannot mate that way! When it comes to courtship there is always the risk that the prospective partner may not be ready, so that a male about to mount a female is best to keep his options open and be all set for a quick retreat if she should attack.

Figure 6.12. The similarity between some display postures and other movements suggests that the former are derived from the latter. Thus stretching and crouching displays shown by many birds (*a*) probably evolved from intention flight movements and some other displays (*b*) show an obvious link with grooming.

This may explain why courting animals show intention movements, but then why should such actions have turned into displays? To do so they would have to signal something useful, that it paid the animal producing them to inform its partner. Here the answer may be that the male is simply telling the female: 'I'm off if you don't do something about it!' If mates are hard to come by it may be worth the female becoming receptive rather than risking the loss of the male.

2. *Displacement activities.* We have already discussed these in Chapter 4: actions which appear out of context to the observer, like grooming in the middle of courtship or nest building in the midst of a fight. They often occur when other actions are thwarted, as when a male is trying to court an unreceptive female, or when the animal is in a conflict. As mentioned above an approach–avoidance conflict may often occur in animal courtship. The occurrence of grooming and bathing movements during courtship are especially common among ducks, the most famous example being in the mandarin, in which the drake has a greatly enlarged secondary wing feather over which he draws his beak when displaying. This action is clearly derived from preening, though the feather is not actually groomed during the display. The structure has also become greatly enlarged to show off the movement to best advantage: a classic example of a releaser.

Figure 6.13. When coyotes threaten each other they strut around with the fur on the back of their neck raised. This is probably a cooling response adapted to serve a display function (photograph by F.J. Camenzind).

Again we can ask what the gain is to the displaying bird in having grooming movements incorporated into its displays. It is not easy to be sure about this but perhaps, again, the male indicating a conflict about whether to approach and mate or to fly away might originally have spurred the female on to becoming receptive lest she lost her partner.

3. *Thermoregulatory movements.* Animals that indulge in strenuous activity, like courtship or fighting, tend to overheat, and various reactions help them to cool down. A coyote strutting stiff-leggedly around its kill, will show its teeth and raise the fur along the back of its neck to any others that appear (Figure 6.13). Showing the teeth obviously displays its weaponry, a show of aggression which hardly needs a complex explanation. Raising the fur helps air to circulate close to the skin and may assist it to cool down in readiness for fighting. It is a very striking sight, however, and there is little doubt that it acts as a signal to others to keep away as well as helping the animal to 'keep cool'.

4. *Protective movements.* Just as aggressive displays often involve showing off the armoury of antlers, horns or teeth that a rival risks encountering, so the animal about to get into a fight will benefit from hiding its more vulnerable parts. It may avert its eyes, or frown so that they are better protected; it may

flatten its ears making them less likely to be torn. All these actions may occur in the heat of a fight, but they also appear when animals threaten each other suggesting that they have also become incorporated into displays indicating that the animal is prepared to attack.

All these aspects of behaviour provide fertile ground for the evolution of displays, because they say something about the state of the animal that shows them. This information in turn may affect the way its partner or its rival behaves towards it. Provided this influence is beneficial to the performer then its behaviour is likely to evolve into a display. Just how then have displays come to differ from the behaviour patterns from which they were derived?

A major difference occurs through the influence of a process that Julian Huxley called ritualisation. If we think about ducks again, when they are bathing they splash up and down, throwing water over their backs, and they beat their wings to wet them thoroughly. The amount of each action, its intensity and the sequence in which the different movements are shown vary enormously. Likewise in grooming, ducks will nibble at the edges of each feather in turn, paying sufficient attention to each to get it back into shape and approaching it from outside or inside the wing presumably according to which best achieves its ends. The equivalent movements in display are quite different, being stylised and stereotyped, the actions often exaggerated but usually very constant in form and in length. The head-toss of the goldeneye drake, referred to in Chapter 2, may well derive from a bathing movement, but it is very fixed in form and in duration, it is exaggerated in that the head is thrown right over till it touches the back and it is stylised in that it does not even serve to wet the plumage. Similar features can be seen in the mandarin's wing preening display. It is rather fixed and constant in form, it is always directed at one particular feather and it does not serve to clean this in any way.

Ritualisation has served to make displays clear and unambiguous so that the observer cannot doubt that the animal is displaying (Figure 6.14). Many displays are remarkably striking and can be seen and heard from long distances away. They have usually ceased to serve the function of the behaviour from which they were derived. Niko Tinbergen referred to them as being emancipated, suggesting that their motivation had also changed from the original. Thus he argued that the preening movement of a courting duck is part of its sexual behaviour system rather than that controlling grooming. It is thus more likely to be influenced by sex hormones than by the amount of dirt on its feathers. Indeed evolution may have gone one stage further than this and led displays derived from intention movements or displacement activities to occur

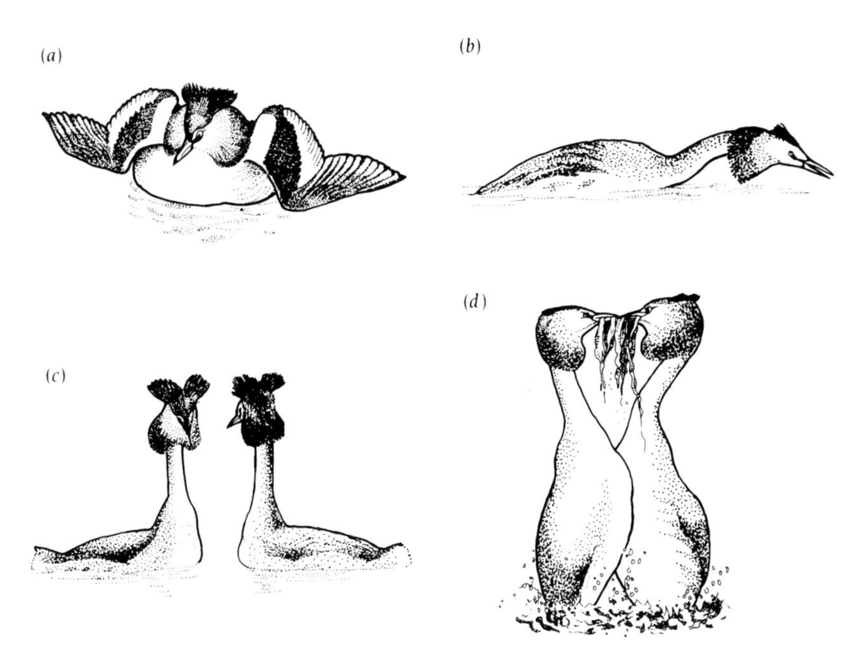

Figure 6.14. Julian Huxley carried out a pioneering study of displays in the great crested grebe during which he developed the idea of ritualisation. Four of the most striking displays are illustrated here: (*a*) The cat display in which the white wing bars are shown off. (*b*) The Dundreary attitude, so named after the long side-whiskers of T. Taylor's fictional character of that name. (*c*) The shaking display, where the members of a pair face each other in a very erect posture and shake their beaks from side to side. (*d*) The weed dance, in which the partners both surface from diving with weed in their beaks and then swim towards each other, rising up and treading water as they come together. (After Huxley, J.S., 1914. *Proc. Zool. Soc. Lond.* **35**, 491–562).

when there is no intention, no conflict and no thwarting, even if these were essential for the display first to appear.

Ritualised behaviour patterns tend to be very constant in form. While people can show a 'slight smile' or a 'broad smile', few animal displays are graded in this way. A cricket chirps or it does not chirp, a fighting fish raises its gill covers in threat or it does not do so. Displays tend not to be half-hearted, but to be all or nothing, appearing at full throttle or not at all. Because of this Desmond Morris referred to them as showing typical intensity, and he argued that they did so because it avoided any possible ambiguity. Thus, when an animal shows a display it is absolutely clear both that it is doing so and also which of its displays it is performing.

John Maynard Smith has pointed out that typical intensity may also have arisen for a very different reason, and this is because it is beneficial to animals to keep their cards rather close to their chests, especially when contesting with each other for resources such as territory. The animal that reveals its true motivation ('This little bit of ground would be quite useful to me but I don't really need it') is giving valuable information to its rival, and the rival can always make a higher bid by bluff: 'I demand this piece of land and am prepared to fight to the death to get it.' Thus negotiations such as these tend to have become highly exaggerated and fixed in form with each side demanding the maximum. There is an amusing parallel here with trade union negotiators threatening all-out strikes unless given vast sums, while managers offer minute rises with the excuse of imminent bankruptcy. Though not, of course, for evolutionary reasons, these demands and offers show typical intensity and are a far cry from the actual moderate figure at which both parties are really prepared to settle!

Courtship and fighting are striking aspects of animal behaviour, which have certainly attracted their fair share of study, leading to the ideas on the origin and evolution of displays that we have just discussed. These ideas are plausible, but the problem of looking back at behaviour in evolutionary time remains a particularly difficult one in this area. There are plenty of hypotheses, but it is not awfully clear how they could be tested.

6.4 Conclusion

Behaviour is a product of evolution like any other characteristic of an animal. Research on behaviour genetics leaves us in no doubt that, despite its great flexibility, behaviour can be modified by selection. But studying its evolution is a more difficult matter. Occasionally, as in the example described in Box 1.3, we can observe it in action, but the timescales involved are usually too long for this. What we are left with is trying to reconstruct the course it must have taken from comparing the behaviour of species alive today. This is easiest with displays, as there are many reasons for them to diverge between close relatives and few to make them converge between distant ones. But there have also been some notable successes in reconstructing the evolution of other behaviour patterns. In doing so we often find similarities arising in ones that are not closely related but face similar challenges, like the kittiwake and the Galapagos swallow-tailed gull. These common solutions to common problems are one way that we get insight into the adaptive significance of behaviour, the subject on which we will concentrate in the next chapter.

7

Function

The word 'function' has many different meanings. Mathematicians manipulate functions, politicians attend them, and students follow courses on how computers function. But to biologists the word has taken on a very specific and precise meaning to which they restrict its use. The function of a feature such as a behaviour pattern is its selective advantage or survival value, the reason why individuals are thought to do better as a result of possessing it. This may seem straightforward enough, but it is a rather easy topic on which to think up ideas (as many popular books about animal behaviour make clear!) and yet a rather difficult one on which to test these ideas and so decide between them. Nevertheless, study of the function of behaviour patterns is a popular and exciting field about which a good deal of firm knowledge has accumulated. Before describing some of these results, we must discuss the concept of function itself in more detail. At the end of the chapter we will go into some of the changes in thinking about how evolution works that have led the ideas surrounding this concept to move to the centre of the stage. These are mainly concerned with social relations between animals, so this will form a basis for the topics of the last two chapters.

7.1 The concept of function

The meaning of function can be best illustrated by taking a particular example. Let us do this by using the question 'Why do birds sing?'. If you put this to several people, you would be likely to get very different replies. Some might say that it was because of sunny weather or because they are male and have the appropriate sex hormones circulating in their blood. These are causal explanations, of the sort we discussed in Chapter 4, rather than the functional ones that concern us here. Other people might say that birds sing to attract a mate or to drive away rivals. These are advantages that song may have and so they are possible functional explanations, based on the consequences of the behaviour rather than on what causes it. These two sorts of answer are not alternatives in any way. All behaviour has both causes and functions, and understanding it fully involves knowing about both. But the two sorts of questions must be kept separate, for one certainly cannot answer causal questions with a functional explanation or vice versa, though people commonly try!

Some functions are easy to understand, without the need for a lot of argument or for complicated experiments. The functional significance of escaping from predators or producing eggs is obvious, for survival and reproduction are clearly of selective advantage, and few would question behaviour that leads to them directly. But there are some actions which are much more difficult to explain in these terms. Perhaps bird song does exclude rivals, but what is the function of that? Less competition over food maybe, and hence more food available for the bird to feed to its young, and hence the possibility of producing more young. Producing more young is without doubt advantageous in most species, as selection generally favours the individual leaving the most surviving offspring. But it takes several links in a chain of argument to get from song to this direct advantage, and other possible lines might be followed instead (Figure 7.1). Only by careful study can one decide which of them is likely to be correct.

An important point here is that behaviour patterns do not necessarily have only one function. Indeed any consequences that lead their possessors to be more successful may help to maintain them in the population. Bird song does exclude rivals, as has been shown by the fact that birds made unable to sing suffer more intrusions on their territories. Song is also known to attract females. In several species they have been found to approach a loudspeaker playing it, and even to flutter their tails up and down in the posture of receptivity normally shown to a male. Finally, a third function of song is to stimulate the reproductive system of females: the ovaries of canaries and budgerigars grow when the birds are played tapes of their own species song, and they will eventually lay eggs without a male being present.

Song, therefore, is known to repel rivals, to attract mates and to stimulate them once attracted. In any one case all these three advantages may be important. But song may also have quite a few disadvantages. A bird singing loudly and obviously on top of a tree must be a sitting target for a hawk; some birds produce literally thousands of songs a day during the breeding season, so they

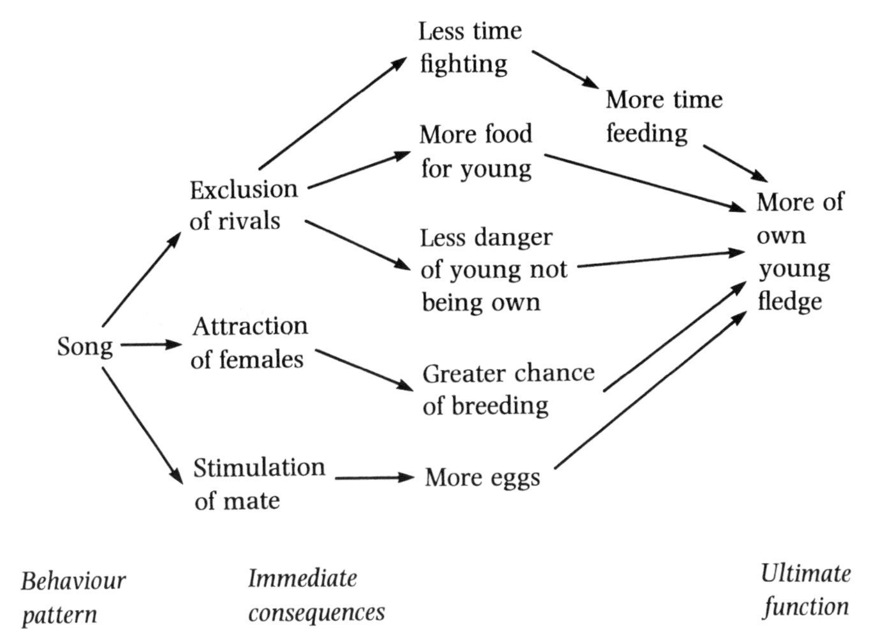

Figure 7.1. Understanding the functional significance of a behaviour pattern involves discovering how it increases transmission of the individual's genes into the next generation. This is the 'ultimate' function of behaviour but, in a case such as song, it cannot be simply linked to the act itself. As the diagram shows, song may have several immediate consequences each of which might lead to greater breeding success. Which of the links of argument is true in a particular case is a matter for study.

are also using up a lot of time that they could be spending feeding or incubating their eggs. The important point is that, for a behaviour pattern like song to be advantageous, the various benefits it confers must outweigh the various costs (Figure 7.2).

The study of function often comes down to looking at behaviour in this way, as if it was a branch of economics, trying to discover just what all its plus and minus points are and, when they are added together, why the answer is positive. But there are several different approaches to examining functional questions: the next section will describe some of them.

7.2 Testing ideas on function

A problem with studying the adaptive significance of behaviour is that it has evolved to function in a particular environment and so it can only be fully

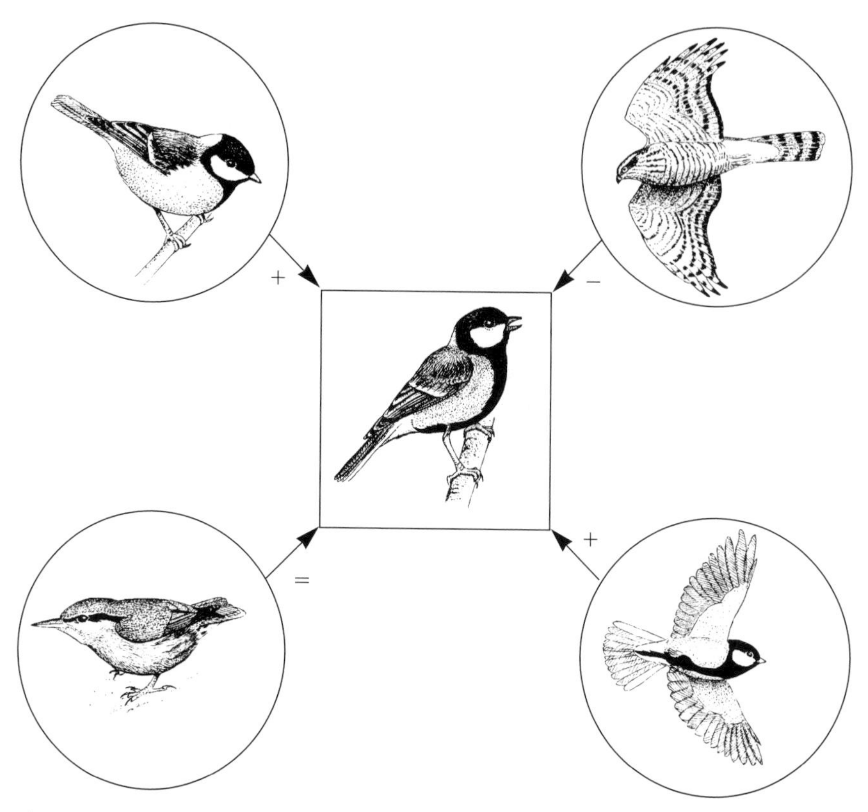

Figure 7.2. Communicating may have both advantages and disadvantages, and the former must outweigh the latter if it is to be favoured by selection. In the example shown here, a male great tit gains if his song attracts a mate (top left) or repels a rival (bottom right), there is no cost or benefit to it being heard by most other species (bottom left), but a disadvantage of singing may be that it attracts predators (top right).

understood in relation to that environment. This raises difficulties, though it does of course give one a good excuse for working in some magnificent places! The main snag is that the usual approach of science, the carrying out of carefully controlled experiments, may be out of the question. There are many features of the outside world, like the weather for example, that cannot be changed at the whim of an experimenter and which may also alter from day to day. Even if the scientist can change a feature and so study the effects of this, it is not often easy to make the change simple and precise so that a comparison can be made between experimental and control groups which differ only in this one very specific way.

Figure 7.3. An adult black-headed gull removes the broken egg-shell shortly after one of its chicks has hatched. Experiments by Tinbergen and his colleagues showed that such broken shells attract predators and thus suggest that the behaviour pattern lowers the likelihood of predation.

As we shall see, some excellent experiments have been done in this area, but much information has also had to come from observation alone where the experimental approach has been impossible. Observations are most fruitful where they also involve comparisons, not in this case between experimental animals and controls, but between different species. That is a topic we dealt with in Section 6.2 where, as well as looking at evolutionary processes, we considered the adaptive significance of the species differences uncovered. The two processes are closely interwoven. But there is also another approach, which is to observe how one species behaves and compare this with how theory predicts that it would be best to behave. We will discuss this in the second section below.

7.2.1 Experiments

The classic example of an experiment on function was carried out by Niko Tinbergen and his colleagues on egg-shell removal in black-headed gulls (Figure 7.3). Shortly after a chick hatches, these birds will remove the broken shell from their nest and carry it away. Tinbergen asked what the function of this behaviour might be. Perhaps the chicks were sometimes cut by the broken edge of the shell if it was left; maybe fluid remaining inside the shell might be a breeding ground for bacteria which could infect them; perhaps predators might be attracted to the chicks and unhatched eggs by the conspicuous white inside of the broken shell. To test this last idea Tinbergen laid out a number of artificial nests containing gulls' eggs, some with broken shells beside them and some not. Returning later he found that the presence of the shell had indeed led to greatly increased egg loss to predators, which could only have been because these were attracted by the broken shell. This suggests that avoidance of predation is one reason why gulls remove empty egg-shells, though this need not be the only reason. He did not test them, so the other ideas Tinbergen had may also be correct.

An interesting point about this behaviour is that the gulls do not normally remove the shell until about an hour after the chick hatches, but remain at the nest during this time while the chick's down dries out. The reason for this appears to be that unguarded chicks are often eaten by other gulls in the colony and they 'slip down' best in the period after hatching when they are still wet. To leave the chick unguarded at this stage would thus risk predation from another source.

We can return to bird song for another good example of an experiment designed to test a functional idea. John Krebs removed all the breeding male great tits from a small wood in which there were eight territories (Figure 7.4). He left some parts of the wood empty, but stocked others with loudspeakers, some of which played tapes of a tune on a tin whistle, while others produced recordings of great tit song. Observation over the next few days revealed that new great tits gradually moved into the wood and set up their territories, but they did not occupy all areas at the same time. The silent part of the wood and that where the whistle tune was relayed were taken over more quickly than the part where songs were being broadcast. This therefore showed that song does indeed act as a 'keep out' signal to male great tits, so that one of its functions is to advertise ownership and so repel rivals.

7.2.2 Optimal animals

Comparing how an animal behaves with how theory predicts that it ought to behave has proved a very useful further approach to studying the function of behaviour. We will consider examples from reproductive behaviour later in the chapter but, at a more day to day level, the most popular subject for study using this approach has been foraging. Many studies have been carried out on the efficiency with which animals feed to see whether this is as great as it could be. The usual assumption here is that the 'optimal forager' should spend as

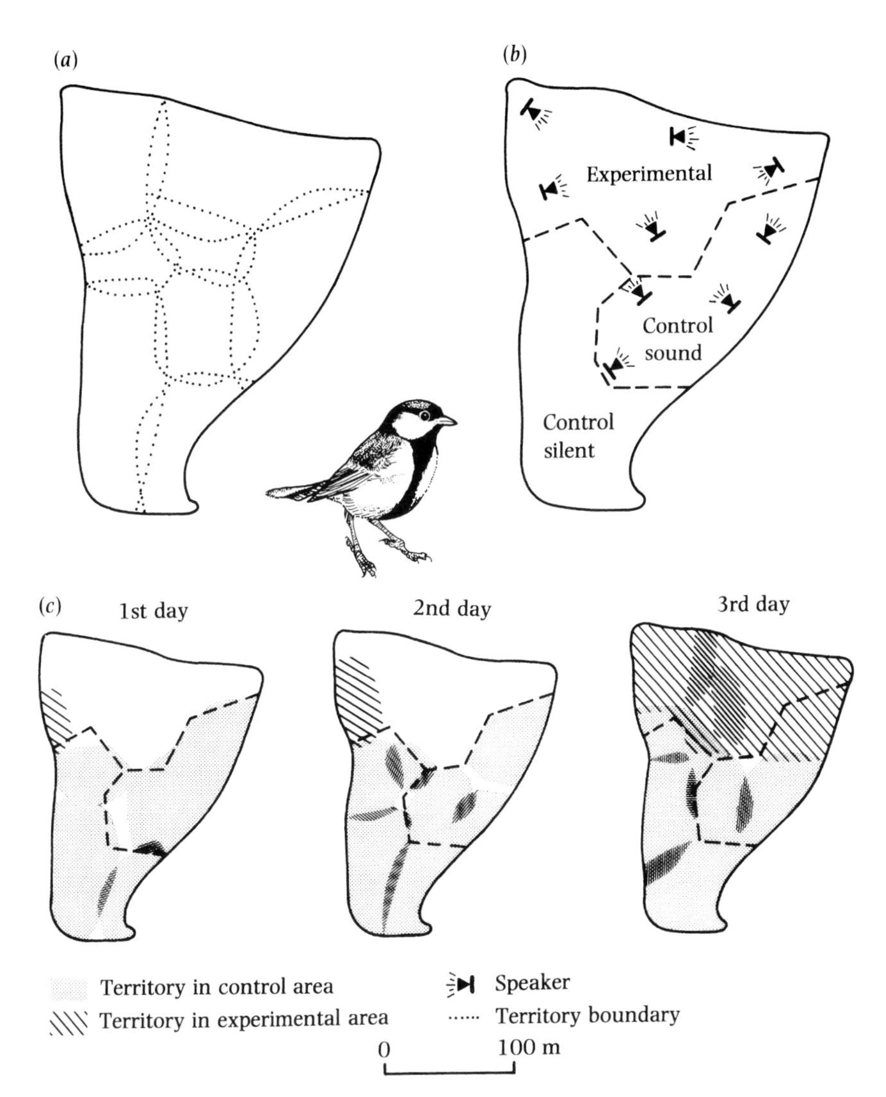

Figure 7.4. An experimental demonstration that song acts to repel rival males in the great tit. All the males were removed from the eight territories shown in (*a*). Three were then 'occupied' by loudspeakers playing great tit song, two by loudspeakers playing a similar phrase on a tin whistle, and three were left silent (*b*). New males settled in the wood as shown in (*c*): the part where song was being played was occupied more slowly than the control areas. (Redrawn from Krebs, J.R., 1977. In *Evolutionary Ecology*, ed. B. Stonehouse & C. Perrins, pp. 47–62. Macmillan, London.)

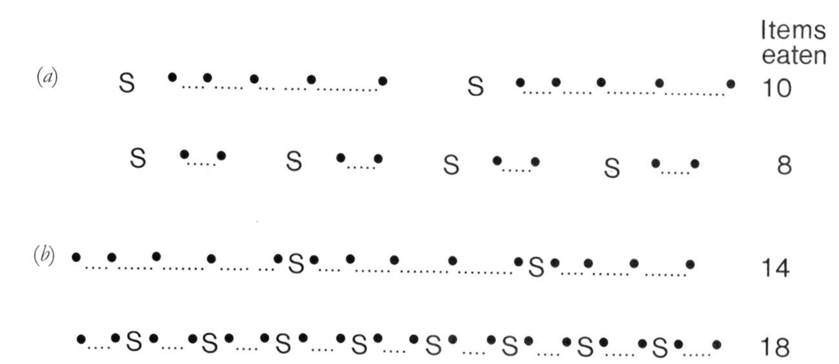

Figure 7.5. The essence of Marginal Value Theorem. Food items often occur in patches for which an animal has to search. In this diagram S represents search time and the black dots are food items. Within a patch these become scarcer as the animal eats them and thus harder to find. In (*a*) patches are far apart or otherwise difficult to find. The animal that does best is the upper one which stays a long time in each patch. In (*b*), however, where patches are close together or obvious, the most profitable behaviour is that of the lower one, which moves between patches long before they are depleted.

little time and energy feeding for the greatest possible returns or, in other words, that it should 'maximise net food intake per unit time'. Generally this assumption holds up rather well when the feeding of animals is studied in detail. We already looked at the example of how bumblebees forage for nectar in Chapter 1 (Box 1.4). Here we will take some examples from foraging birds. Several different sorts of question arise.

Where to look for food A lot of laboratory experiments have been carried out on this subject, and they point to animals following some quite sophisticated rules. For example, when an animal feeds in a small area, the food will become progressively depleted and individual items more difficult to find. When should it move elsewhere? To maximise food intake, theory suggests that this should depend on how far away the next patch of food is (Figure 7.5). If it is close, the animal should move earlier so that it will benefit from the high intake level it will achieve when it gets there. But if it is far away the animal is better to spend longer in the present patch, rather than making a long journey during which no food at all will be obtained. Laboratory experiments on tits suggest that they do indeed do this, their behaviour fitting in with this so-called Marginal Value Theorem.

Animals may also have to decide between areas that yield different foods. Nick Davies found that spotted flycatchers varied between searching in the

Figure 7.6. The two feeding techniques of oystercatchers penetrating mussels, hammering and stabbing.

canopy for aphids and sitting on low perches waiting for large insects such as bluebottles to come by. In the middle of the day large insects become much more active and the latter strategy is very profitable – far more so than gleaning aphids could ever be. But early and late, when larger insects are scarce, the aphid strategy is the more profitable of the two, and this is the one that the flycatchers follow.

What prey to take There are a number of aspects to this question. Prey size is perhaps the simplest. If prey is scarce, birds tend to eat everything that comes their way. But if it is common they do better if they ignore small prey, which yield less profit for the amount of time it takes to handle them, and just concentrate on larger ones. Several studies have shown that this is indeed what they do. A neat solution is found in the case of some birds which forage for their nestlings and so must carry items back to the nest, a situation known as central place foraging. In wheatears, for instance, the adult will move away from the nest amassing small items as it goes until it has a beakful to take back to the chicks. But if it comes across a large item, which would itself amount to a full load, it swallows any earlier prey it has caught, and flies back with the big one.

Just as bumblebees develop the skills appropriate to different flowers, so many birds benefit from concentrating on foods or techniques at which they are best. Oystercatchers, despite their name, feed largely on mussels. Some birds prefer 'stabbing', inserting their beak between the valves and snipping the muscles that hold them closed. But other birds specialise in 'hammering', cracking the valves themselves to gain access (Figure 7.6). Within this group birds even specialise further, in which valve they crack and which area of it they go for. They become adept at choosing fast growing mussels, which have thinner shells and which they may be able to identify because they tend to be browner and to have fewer barnacles growing on them.

What route to follow Bernd Heinrich found that the bumblebees he studied (Box 1.4) moved round the area in which they live in a systematic fashion, visiting each clump of flowers in turn rather than going back to one they have fed on recently. This enables the flower to replenish its nectar supply and means that the insect gets a reasonable reward for each visit. Birds often also 'maximise return time' so that food supplies are likely to have replenished by the time they come back again. Jamie Smith found that thrushes feeding on a lawn did so by moving in a series of short runs, between which they scanned. Unless the bird found food, the angles between successive runs tended to be small and to alternate, so that it zig-zagged across the lawn and covered a lot of ground. However, if it found food, such as an earthworm, its subsequent moves were usually shorter, with the angles between them sharper and all in the same direction. This had the effect of keeping the bird in the same area, an adaptive response as worms tend to be clumped, so there is a good chance that a second would be found in the same area as the first.

What rules to use The idea of applying optimality theory is not to see whether animals behave in an optimal fashion, but to examine the rules that they use. In all the bird examples given above, they appeared to be complying with the most commonly tested rule, that they should behave so as to 'maximise net food intake per unit time'. But there are circumstances in which other rules apply. For example, the foraging of some animals is risk sensitive. When very hungry, rather than feeding in an area where the food is scarce but evenly distributed, they go for one where it is more patchy, but there is at least a chance that they will hit a good patch. If their back is really to the wall, the latter may be the only possibility they have of survival. In social insects another rule may be to maximise information transfer to the colony. Leaf cutter ants finding a good source of food take less of it back to the colony, but run faster and recruit other foragers to visit the source more actively when they get there. As individuals they do not maximise their returns per unit time, but in the long run the colony does better as a result of this strategy and this benefits all its members.

That many animals should have been found to forage in as efficient way as they could is, in some ways, rather surprising. There are several reasons why we might expect them to fall short in this respect. For example, energy intake may not be what matters to them, for there may be all sorts of other factors that will influence the way they behave, from the need to find minor nutrients to the necessity of watching out for predators. Another important influence may be whether or not we are observing them in the situation to which evolution has adapted them. In the laboratory we certainly are not doing so but, even in the wild, the environment has been so modified by human activities that one should not always expect to find animals which are beautifully adapted to its present state. But studies of foraging behaviour do certainly show an impressive capacity on the part of many animals to organise their behaviour in the most profitable fashion.

Throughout this section various ideas about what natural selection has led animals to do have been mentioned. In talking about foraging we assume that efficient animals do better than inefficient ones and that selection will favour the former. Earlier we discussed other ways in which animals enhance their own survival and reproductive success, on the assumption that this was to their advantage. Now, however, it is time to look at these assumptions a little more closely, for there has not always been agreement among biologists about what it is best for animals to do.

7.3 Selfish genes

The current interest in the function and evolution of behaviour stems from a synthesis between animal behaviour and evolutionary theory which had curious and rather inauspicious origins nearly 40 years ago. Something of this history will help to put in context the growth of this subject, which is often now referred to as behavioural ecology.

In 1962, V. C. Wynne-Edwards produced an impressive book called *Animal Dispersion in Relation to Social Behaviour* in which he suggested that the behaviour of animals can be best understood if they are viewed as acting for the good of the species or group to which they belong. He proposed that many communal displays, ranging from bird roosts to mosquito swarms and the vertical migration of plankton, allow animals to assess population density and so adjust population size to match the food available. He also argued that animals do not fight very frequently because harming others would do damage to the group. On the other hand, they may share food with each other because this is to the group's advantage. If food is scarce some of them may even forgo breeding so that the size of the group does not reach a level at which individuals would start to die of starvation.

Wynne-Edwards simply made clear the assumptions and thinking of many biologists at that time. The value of his contribution was not that it converted others, but that it caused them to look at these ideas in more detail and ultimately to reject them. If members of a group do not breed because food is running short and the group might suffer if they did, what would happen, some people asked, if a mutant animal arose that failed to obey this rule. The answer is simple: it would pass more of its disobeying genes to the next generation so that these would spread through the group as generations went by. Eventually all animals in the group would breed as much as they could, regardless of whether the group as a whole would benefit from restraint.

Points such as this led to acceptance of the idea that selection acts primarily on individual animals, favouring those that do better in the business of passing on their genes at the expense of those that do worse. The term 'fitness' is often used in this context to describe the capacity of different individuals to contribute to the next generation; those of greater fitness make a larger contribution so that their genes tend to spread. Such expressions as 'acting for the good of the species' have, as a result, dropped almost totally from use. This change in thinking has strong implications for behaviour, and especially for the social relationships between animals. But how can one explain the huge number of examples that Wynne-Edwards amassed to support his argument, if its very basis is unsound?

Aggression becomes easier to understand if individuals act only for their own good, indeed one might expect them to fight a tremendous amount the whole time, each being out for its own ends and careless about possible damage to others. This certainly does not occur, but the reason is probably simply just that fighting is dangerous. Even if one animal is much bigger than another, the small one may give it a scratch that will go septic or a fatal nick in the jugular vein before it is subdued. Although the risk of injury may be slight, it will not be worth taking unless the fight is over something very valuable. Nevertheless fights to the death have often been described among animals. When thorough field studies have been carried out, for example someone has watched a small troop of monkeys for a few thousand hours, it has not been unusual for one animal to be seen killing another. It would be quite exceptional to observe such a high rate of murder in a human society, despite the supposedly high aggressiveness of our species!

A good example here, and one which certainly would be hard to fit in with thinking in terms of the good of the species, is what occurs when male lions take over a pride (Figure 7.7). A pride normally consists of a small number of males, females and their cubs. Some males also go around in bachelor groups that wander over the savannah looking for prides to take over. This they are likely to succeed in doing if they come across a pride in which the adult males are few, aged or out of condition. There will be a scrap, at the end of which

Figure 7.7. Young male lions in groups attempt to take over prides from the males in them. If successful the new males slaughter the dependent young fathered by these earlier males and this makes the females receptive again and so brings forward the birth of their own young. (From *Behaviour and Evolution*, ed. P.J.B. Slater & T.R. Halliday. Cambridge University Press 1994. © Priscilla Barrett.)

the previous males are driven off. One of the first things that the new males do thereafter is to kill the cubs already in the group, behaviour one might think of as wanton cruelty. It is, however, easy to see how it benefits the males themselves. They may only be able to control the pride for a short time and the more young they have during that period the more of their genes they will be passing on. By killing cubs that they did not father, they make the mothers stop lactating and become receptive and thus they speed on the time when these females will bear cubs sired by themselves. Indeed, time is of the essence, for if these cubs are not well grown by the time their fathers in turn get ejected, they too will be slaughtered and the whole investment wasted. In this way infanticide, which also occurs in many other species, can be understood as of advantage to the individual, though it would certainly not be to the group or the species.

Aggression in animal groups is, therefore, not very rare. It tends to be limited where it could have a dangerous outcome for either of the participants: in this case it is commoner for animals to display and threaten each other until one backs down, an aspect of communication to which we shall return at the end of this chapter. But what then of altruism? Surely selfish animals should not share bananas, and yet chimpanzees often do this. The young of many bird species, from European long-tailed tits to Arabian babblers and Florida scrub jays, remain with their parents and help to feed later broods rather than going off to have families of their own (Figure 7.8). Most

Figure 7.8. Helpers at the nest: in many bird species, other individuals help the parents to feed their chicks. In most cases these helpers are the young of previous broods. This photograph shows a mated pair of red-throated bee-eaters and (in the middle) their son, hatched in the previous year and now helping at the nest (photograph by C.H. Fry).

remarkably, some social insects like ants and honey bees have workers that are completely sterile so that they never breed themselves. They have sacrificed all their own fertility and devote themselves selflessly to raising the offspring of others in the colony. Usually, as with helpers at the nest in birds, the young ones that they raise are their own brothers and sisters.

A major contribution to our knowledge of how evolution works, which helps us to see how such behaviour could have arisen, came in two papers by William Hamilton published in 1964. He took as his basis the point that what really matters as far as natural selection is concerned is the extent to which genes are passed on from one generation to the next. Genes leading animals to behave in ways that enhance this transmission will spread at the expense of others. Now animals producing more young than their rivals are clearly succeeding in this way. But what of those that help their parents, their sisters or their cousins? Relatives share genes, to a diminishing extent as the relationship gets weaker, so assisting close kin may actually help an animal to get more of its genes into the next generation, though in a more roundabout way than by

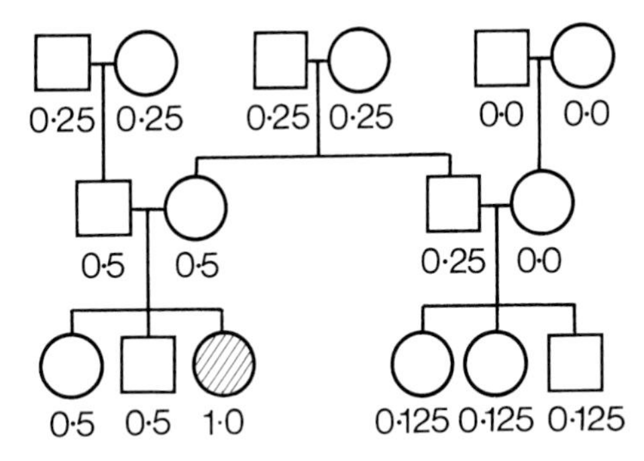

Figure 7.9. A pedigree diagram illustrating relationships within a family. Squares and circles indicate males and females respectively. The figures are coefficients of relatedness (r) between the female (shaded) and all the other individuals. Within the diagram she has two siblings, two parents, four grandparents, an uncle and three first cousins. (Adapted from Shields, W.M., 1984. *Philopatry, Inbreeding and the Evolution of Sex.* State University of New York Press, Albany.)

breeding itself. This theory is referred to as that of 'kin selection' because it explains why animals are prepared to behave in ways that involve sacrifice on their own part but advantage to their kin. The argument put forward by Hamilton is that animals should not just maximise their own fitness but what he refers to as their 'inclusive fitness'. This has both a direct component from their own reproductive efforts and an indirect one from the help they give to relatives. A young bird which stays to help its parents with the next brood will therefore be making a contribution to its own inclusive fitness. If breeding requires skills it may also be gaining useful practice and, if good territories are few and far between, it may have more success in the short term by staying at home than by going off and trying to find one of its own. In the long run it may even achieve 'promotion' by becoming a breeding adult itself once its older relatives have died. Though it appears at first sight to be selfless altruism, helping at the nest may therefore be very much in the bird's own interests and not, strictly speaking, altruism at all.

The extent to which it pays an animal to help a relative depends on how closely related the two are. To assess this, we use a measure called the Coefficient of Relatedness, or r. This varies from 0 to 1 and is best viewed as the likelihood that two individuals share a particular gene through their shared inheritance (Figure 7.9). If they are identical twins, they are bound to do so, and $r = 1$. If they are totally unrelated $r = 0$. For brothers and sisters, or a parent and its offspring, $r = 0.5$. In the case of full cousins $r = 0.125$, and so on. These figures help one to realise why helpers at the nest are so common:

a young animal is as closely related to a full brother or sister as it would be to a son or daughter. In terms of its genes, it is therefore as beneficial for it to rear a sibling as one of its own young.

In his 1964 papers, William Hamilton worked out a general rule: it would benefit an animal's inclusive fitness to help a relative provided that:

$$b/c > 1/r$$

where b is the benefit to the relative, and c is the cost to the helper. So, as long as the benefit is greater than the cost, it pays to help an identical twin (or yourself!). On the other hand, the benefit must be huge in relation to the cost for an animal to go to the assistance of a third cousin twice removed.

Aggression and altruism are just two aspects of behaviour that we can understand better when we start to think in terms of animals acting so as to maximise their inclusive fitness. But, for most animals, their major contribution to the next generation is through their own reproductive efforts. In jargon terms, they are maximising the direct component of their inclusive fitness. In the next section we will consider various ways in which this is achieved.

7.4 Strategies of reproduction

Animals that mate with one another are not usually close relatives, as inbreeding often leads to weaker offspring and so tends to be disadvantageous. Nevertheless, sexual partners have a shared interest in the young that they produce and so they often cooperate in raising them. Indeed one might think they should always do so to maximise the chances of the young surviving. But this is far from so. Many birds, and a few mammals such as marmosets, gibbons and some humans, are monogamous, forming stable mated pairs the two members of which combine to rear their young. But there are several other ways of doing things. Toads, for example, produce hundreds of eggs and leave them to fend for themselves. In ducks, the males appear to have an easy time of it: they mate and then go off to moult in stag parties, while their females lay, incubate and raise the young unaided. Sticklebacks, as we saw in Chapter 1, are the opposite. The male builds the nest and tends the eggs and young, the female contributes only her eggs.

One can begin to understand why different species adopt different strategies if one looks at just what the costs and benefits of each way of doing things may be. Various theories, some quite complicated and mathematical, have been devised with this end in mind, especially by Robert Trivers and others influenced by him. Here we can do little more than give some impression of the reasoning behind them. Young which taste nasty, like toad tadpoles, and which can fend for themselves as easily whether their parents are around or not, might as well be left. The parents will leave more descendants by putting all their energies into eggs and sperm and none into caring for them. But more vulnerable offspring have no chance of survival without parental care and, indeed, producing few young that are then carefully tended may be much more productive than having lots that are left to their own devices. Just how they are cared for depends on the advantages to each of the two parents. If both together can rear twice as many young as one alone could, or if one of them leaving its partner to tend the brood has little chance of finding another mate, then both are likely to help. On the other hand, if one parent can do quite well on its own, then it may pay its partner to desert. The one that deserts is, of course, behaving in a 'selfish' manner but, if it gets more genes into the next generation as a result, natural selection will favour it deserting.

Which partner is likely to be the one that deserts? The answer here seems to be that it will be the one most likely to be able to breed again easily. If there are a lot of unattached females around, a deserting male may well find a new partner and, provided he does not have to expend weeks of effort in getting another territory, defending it and building a nest before his second breeding effort pays off, it will be worth his while to have a series of mates rather than sticking to one. A good example here concerns the pied flycatcher, a small European songbird (Figure 7.10). These birds nest in holes in trees, or in nest boxes, the female incubating the eggs and both partners sharing in care of the young. In this circumstance they raise, on average, about five chicks. Sometimes, however, a male will move off while his female is sitting and occupy a new hole some territories away with a second female. After this female has started to incubate he then returns to help his first one care for their brood. The second female is therefore deserted and must rear her young alone; nevertheless she usually succeeds in fledging over three, so she does gain more than if she lacked a partner altogether. It is even possible that she may not be able to avoid her situation, if she cannot tell whether a male has another mate elsewhere or not, although this is doubtful as males on their second territories do behave differently. By contrast, the first female has the same number of young as she would do were her male not bigamous, for she still has his help in feeding them. The real gain, however, is to the male, who fathers eight or more surviving offspring, far more than he could achieve with only a single partner.

Females too may move from one partner to the next, though this is less common. Sticklebacks do it, perhaps because there are abundant males and

	Total success of female	Total success of male
Young raised by monogamous pair	5.62	5.62
Young raised by 1st female with bigamous male	5.73	9.33
Young raised by 2nd females with bigamous male	3.60	

Figure 7.10. In a wood occupied by pied flycatchers, males may defend two territories, attracting a female to the first and then leaving her to incubate while he attracts another female to his second nest box. He does not help this second female to rear the young but returns to help his first. The table shows how the male gains strongly from this strategy and the first female does as well as if he were monogamous, but the second female is unable to rear as many chicks. (Data from Alatalo, R.V. & Lundberg, A., 1990. *Adv. Study Behav.* **19**, 1–27.)

because after the huge effort of egg-laying the female would have little energy for parental care. She may thus gain more by leaving the male to care and, if she can manage to do so before the breeding season ends, feeding so as to build up a further supply of eggs that she can leave with a second male. Generally, however, desertion by the female is rare for several reasons. One is that eggs are a very large investment and it takes a lot of time and energy for the female to replenish her supply, while males can produce huge numbers of sperm at frequent intervals. They can therefore go from mate to mate fathering large numbers of offspring, while the gain in changing mates to a female is much less. A second reason is that females are often literally tied to their offspring. A female mammal can hardly leave her young to fend for themselves or be looked after by their father in the middle of pregnancy! Nor indeed can she do so soon afterwards for they continue to be dependent on her for milk. A third point is that females have more certainty that their young are their own than do males, for it is seldom possible for males to guard their mates assiduously enough to be sure that they have not mated with others, while a female can be certain that an egg she laid or a young one she produced is her own. In other words, males suffer from 'paternity uncertainty'.

In many species, paternity uncertainty is a real influence, for mating with

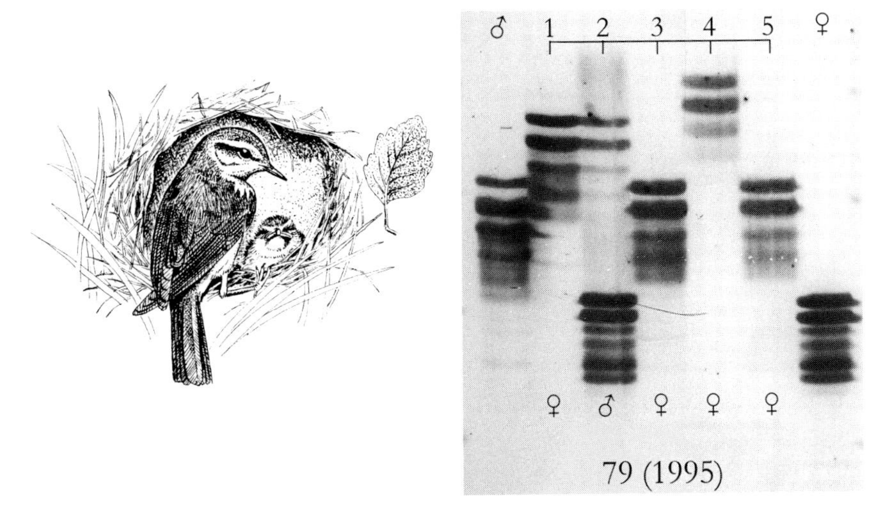

Figure 7.11. An example of DNA fingerprinting showing the minisatellite banding pattern of each chick in a willow warbler nest compared with that of the birds feeding them. In this example (a sex-linked one) males have two sets of bands, but females only one, though the two sets in the adult male in this case are superimposed. A female chick gets these bands from her father. The match between them suggests that chicks 3 and 5 are daughters of the pair male, while chicks 1 and 4 must have another father. Chick 2, the only male in the brood, is the son of the female at the nest, but not the male, though he may share a father with chick 1. Only two out of the five chicks in this nest were thus fathered by the male feeding them (DNA fingerprint kindly donated by Diego Gil).

another's partner while he is off feeding will certainly do wonders for a male's inclusive fitness: he can then help his own partner to raise their shared offspring while his cuckolded neighbour raises other young for him as well. Such 'extra-pair' matings have been observed in many species, but only recently has it been possible to assess their influence. The technique of DNA fingerprinting, famous for its notable successes in linking murderers to their crimes on the basis of minute blood samples, can also be used to look at relatedness. Thus, with a sample of blood from both adults of a pair and all the chicks in their nest, the sharing in the fingerprinting pattern can be used to tell who is related to whom (Figure 7.11). Usually most of the chicks share part of their pattern with the male and part with the female, showing that they are indeed their offspring. But some may only share with the female, indicating that the male feeding them is not their father, and some may share with neither because the egg from which they hatched was dumped in the nest by another pair. Studies like these have shown that extra-pair paternity varies a lot between species, but can be very common. In reed buntings it has been found in 86% of nests and involves 55% of chicks. Nests even occur where none of the young was fathered by the male that is feeding them.

These examples from reproduction show just how powerfully modern evolutionary theory can shed light on various aspects of the behaviour of animals. Many of its facets, which previously appeared hard to account for, have come to seem obvious with the help of the new body of theory. The notion that animals behave in such a way as to maximise their inclusive fitness is a powerful and persuasive one.

7.5 Evolutionarily stable strategies

The cumbersome phrase that heads this section describes one final theory that has done a great deal to shed light on animal behaviour in the recent past. The idea of evolutionarily stable strategies, usually shortened to ESS, was developed by John Maynard Smith. It is based on a branch of mathematics called game theory, which can be applied to any situation (like a game) in which the best strategy for one individual to adopt depends on what others are doing.

In most species all individuals of one sex adopt a similar reproductive strategy, and thinking about costs and benefits may suggest to us why they follow this one rather than any of the other possibilities. We might conclude that no animal that happened to adopt another strategy could do better than if it conformed. There is in this case a single evolutionarily stable strategy which it benefits all individuals to follow. In other cases, however, a mixed ESS can occur and it pays some individuals to do one thing and others another. The foraging of bumblebees is a good case here, though not from reproduction: the flowers it is best for a bee to visit are those being least exploited by others, so the bees end up with different strategies from each other but all equally productive. The reproductive behaviour of male green tree frogs provides another instance. These animals form assemblies beside ponds and set up calling choruses to which females are attracted. However, calling is dangerous and energy consuming, as well as making it hard to keep a look out for the females that approach. Some males avoid these disadvantages by adopting 'satellite' status instead of calling (Figure 7.12). They sit in wait near calling males and watch out for females, intercepting them as they arrive and often succeeding in mating with them before they reach the caller. Not all males can become satellites, however, because then the pond would be silent and no females would come, so the stable situation is a population consisting of a mixture of the two sorts of frog.

Figure 7.12. Male green tree frogs gather on the edge of a pond and call to attract females. However, satellite males that do not call lurk round the edge of the group and may succeed in intercepting females that approach and so achieve mating without incurring the cost of calling. The lower of the two males shown here is calling, while that above is a satellite that is also looking out for approaching females (photograph courtesy of H.C. Gerhardt).

In frogs a male may be a caller one night and a satellite the next. In some species, however, individuals are committed to one strategy or another. A good example here is in the marine isopod *Paracerceis* which lives in sponges. There are three separate male morphs. The largest males belong to the α morph, and these defend harems of females. The next largest are β males, and these behave and look like females, and so enter the harems undetected and fertilise the females in them. The smallest males are γ individuals and these hide in the recesses of the sponge out of reach of the harem males, sneaking out to achieve matings. Observations on this system by Stephen Schuster and Michael Wade suggest that it is a mixed ESS, the three strategies being in the long term equally successful, depending on how many females there are. If there is only one, the α does best, but if several the β individuals may sire at least 60% of the young, and if there are a great many females the γ males do better still. Taking into account changes in female numbers over two years, pointed to the three sorts of male leaving about the same numbers of offspring.

ESS theory can be applied to any situation where different strategies are possible. It sometimes suggests that several should coexist, but in other cases one pure strategy is best and should sweep the board. In its original application Maynard Smith used it to look at fighting in animals to try to understand why they sometimes display at each other rather than attacking. He pitted three strategies against each other: hawk (attack at once and only retreat if injured), dove (display and retreat if attacked) and retaliator (display but attack if the opponent attacks you). This involved some algebra, or computer simulations, with various assumptions about the costs and benefits of each strategy. But the most successful one turned out to be retaliator, which may account for why this is the strategy most often seen among fighting animals in nature.

The theory of ESSs thus sheds light on a variety of behavioural problems, which were difficult to explain before it was proposed. A final one, which we will discuss in Chapter 9, is the question of why unrelated animals should cooperate with one another.

7.6 Conclusion

In looking at function we have to think of how various actions have come to be favoured by natural selection. It is now clear that survival and reproduction are not even all there is to this matter. Indeed survival is of no consequence in itself. It is the passage of genes from one generation to the next that matters, and long-lived animals do not necessarily have greater success in this. The revolution in our understanding of how evolution works over the past few decades has led to extensive study of animal behaviour and a rethinking of many old examples in terms, not of how particular actions benefit the species nor, strictly speaking, how they may be of advantage to the individual's own reproductive success, but of the ways in which every animal behaves so as to maximise its inclusive fitness.

In the next two chapters we will return repeatedly to ideas such as these from modern evolution theory for, especially in communication and the social relationships between animals, these new theories have done a great deal to illuminate the reasons why animals behave in the ways that they do.

8

Communication

Many of the examples discussed throughout this book have involved communication between animals, for this has always been a subject of particular interest to those studying animal behaviour. Indeed the theories that have been developed have in some ways been biased by this interest. Acts of communication do tend to be very fixed in form, to act as releasers for other actions, and to develop very similarly in all members of a species. Perhaps ideas such as that of the fixed action pattern or the innate releasing mechanism would have come less easily to mind if ethologists had been more interested in grooming or feeding behaviour than in aggressive or courtship displays. But then they had every reason to be enthusiastic about studying the latter, because the signals that animals produce in communication are often remarkably striking. Discovering just what they are saying to each other is one of the most fascinating fields of behaviour.

Whenever we see an animal that is brightly coloured or boldly patterned, or which is making a loud and obvious noise, we can be pretty certain that communication of some sort is taking place. Standing out from the background is dangerous and there must be some benefit to doing so which offsets the risks involved. Most often the communication is between members of the same species, and the risks come when predators cue in on this as a means of finding a meal. Sometimes, however, different species communicate with one another and here it is often prey animals that forsake their camouflage and communicate directly with predators that might eat them. These different sorts of relationship between predators and their prey make a useful starting point as they illustrate some basic principles of animal communication.

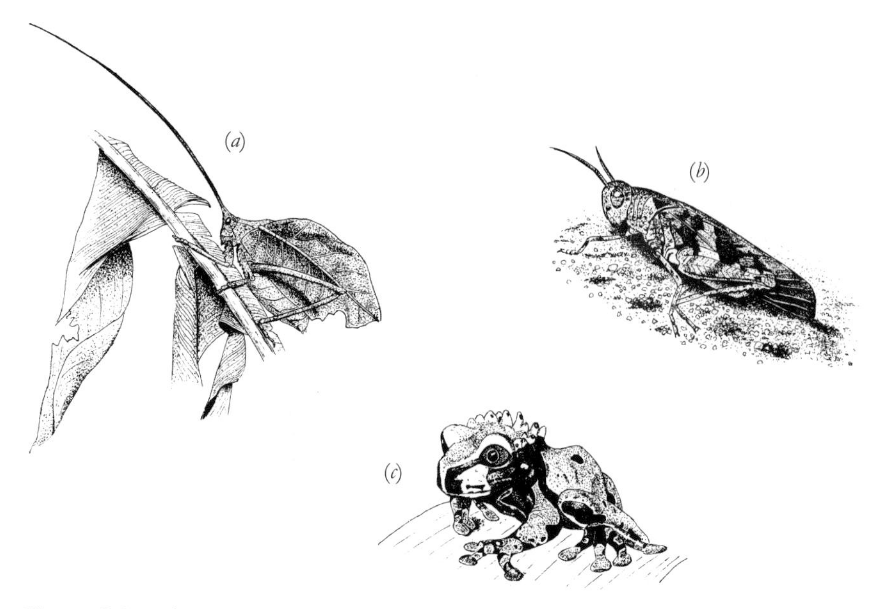

Figure 8.1. The patterning of most animals makes them merge with their background and some of them are remarkably well adapted in this respect: a katydid that mimics a dead leaf (*a*); the grasshopper *Humbe tenicornis* from Kenya with its disruptive pattern and coloration which matches it well to the varied sand background on which it lives (*b*); the crowned tree frog of Costa Rica with markings which make it mimic a bird dropping.

8.1 Predators and prey

Normally, animals are as cryptic and silent as they can be, and it is amazing the accuracy of camouflage that some of them have achieved (Figure 8.1). Selection quickly removes those that fail to comply: the slightest rustle on a woodland floor is enough for a barn owl to home in with deadly accuracy on the mouse that made the sound.

As with other behaviour patterns, then, communication has costs as well as advantages. The animal that advertises for a mate could well end up being eaten instead. This is nowhere better illustrated than in the túngara frog from Panama in which the males call in choruses to attract mates. There is a problem, however: fringe-lipped bats home in on the calls and can catch and feed on the frogs with great success (Figure 8.2). The frogs cannot obtain mates unless they call, so they must do so, but the intense selection pressure that the bats provide has made them minimise their risk. On all but the darkest nights, they spot approaching bats and, while they often ignore the smaller

Figure 8.2. A fringe-lipped bat, attracted by the chorus of túngara frogs around a pond, swoops in to capture one of the callers. (From *Behaviour and Evolution*, ed. P.J.B. Slater & T.R. Halliday. Cambridge University Press, 1994 © Priscilla Barrett.)

species, their chorus will shut down the minute a fringe-lipped one appears above their pond.

Another good example of communicating when a predator is around, at the same time as minimising the risk of being captured, is the high pitched and thin 'seeep' alarm call that many small European songbirds produce when they spot a hawk (Figure 8.3). Interestingly, the call is almost identical in all the species that produce it. Two factors have probably contributed to this. First, there is no need for the call to indicate the species of the caller. This is irrelevant, provided that its mate or young gather the information that there is a hawk about and so know that they should beware. Thus there is no reason for the calls to be different. Second, there is good reason for the calls to be the same. This particular sound has characteristics that make it especially difficult to localise, so all species seem to have homed in on the same form to reduce the risk of capture as much as possible. Animals usually locate sounds by detecting differences between their ears: for example, if a sound is louder in one ear than the other it obviously comes from the side of the head where it is most intense. However, with thin, high pitched sounds, like the 'seeep' call, such differences are slight and locating the caller becomes very difficult for a predator. Certainly, to a human, the call is almost ventriloqual and it is very hard when hearing it to decide where the caller is perched. In the case of

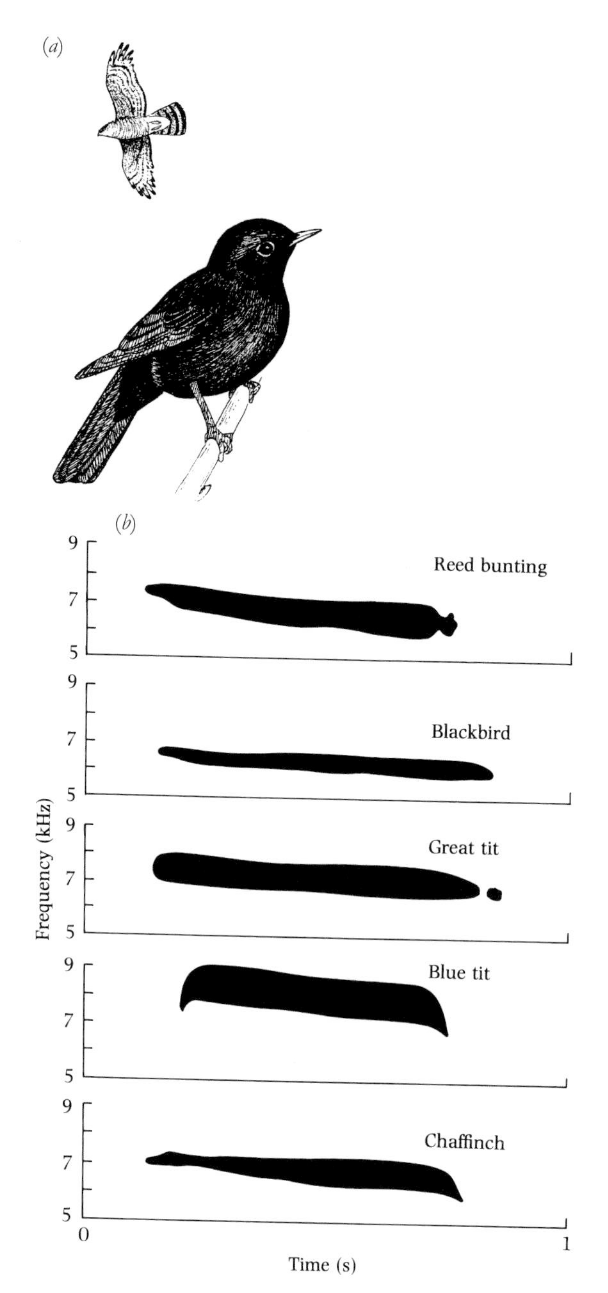

Figure 8.3. (*a*) A blackbird freezes on its perch and produces a 'seeep' call as a sparrowhawk passes over. (*b*) Sonagrams of these calls as produced by five different species showing that they are almost identical. (*b* after Thorpe, W.H., 1961. *Bird Song*. Cambridge University Press, London.)

Figure 8.4. A nightjar lures a fox away from her chick with a broken-wing distraction display. This signal benefits the bird but is to the disadvantage of the fox as it is deceived into moving away from an area where it would find an easy meal.

hawks there may even be an added advantage of calling at this high frequency as they may literally be unable to hear the sound: their ability to hear is much worse at that level than that of small birds.

These examples illustrate how pressure from predators may mould the form that animal signals take, assuming that most animals avoid communicating with predators as far as possible. But what about cases where signals are thought to be produced specifically to communicate with predators? Very often the information that these provide is misleading. Such signals are usually referred to as deceitful, without of course implying that the animals producing them are conscious of lying in any way: they make such signals simply because selection favours the transmission of false information. The bird that drags its wing along the ground as if injured and so distracts a fox from the area of its nest (Figure 8.4) is communicating with the fox, but the information it is providing is false. Its wing is not really broken and, once it has lured the fox far enough away for its chicks to be safe, it will fly off. Likewise, the moth that flicks its forewings forwards to reveal two large eyespots deceives the bird about to eat it by mimicking a large animal with its eyes spaced wide apart (Figure 8.5). Insects which taste nasty or have stings often advertise their distastefulness by bright colouration so that predators soon learn that they are to be avoided. In this case both predator and prey gain: the prey escapes damage and the predator avoids an unpalatable meal. Other species have, however, evolved so as to mimic the unpleasant prey while being perfectly palatable themselves. The black and yellow stripes of wasps and bees warn birds of their stings; hoverflies are harmless but similarly striped and they may thus avoid capture without the need to be dangerous in any way.

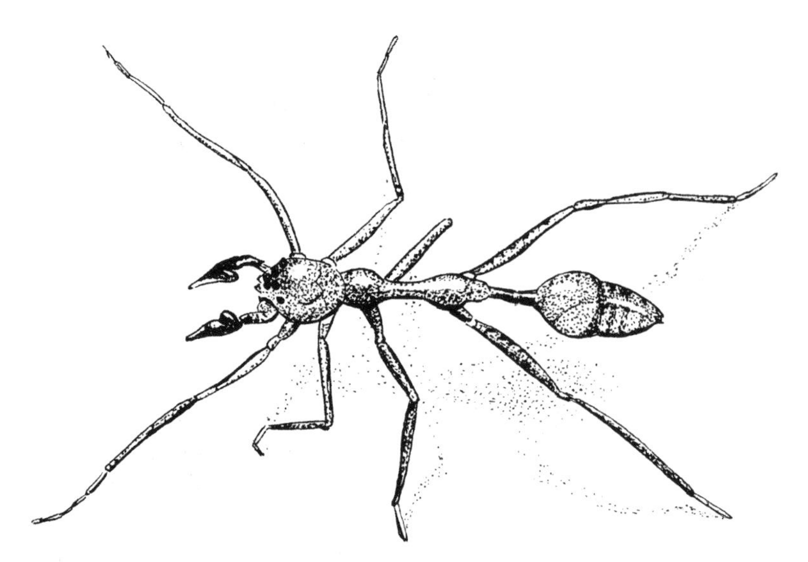

Figure 8.5. A fine case of ant-mimicry by the spider *Myrmecium* from Trinidad. The animal appears to have only six legs like an insect because its front pair are waved in the air as if they were antennae. Its narrow body, and jaw-like pedipalps complete the illusion.

Prey animals can thus warn predators to keep clear, or deceive them into doing so under false pretences. Predators can behave in a similar way, in this case appearing to be harmless when they are actually lethal. The polar bear is white so that it blends in with the snow and can creep up on its prey unseen. An American buzzard, the zone-tailed hawk, which preys on small mammals, also merges with its background, but it does so by being just like a vulture in silhouette. Prey animals pay little attention to vultures as they eat only carrion, so the buzzard gains from this deception by getting close enough to pounce without them running off. The green colour and rather leaf-like limbs of preying mantises probably also help them to sneak up on prey unseen. These examples are of predators being camouflaged. However, predators too may produce signals which mislead their prey, so-called aggressive mimicry. An especially nasty example is that of the firefly *Photuris* (Figure 8.6). Male fireflies produce a signal of light flashes in a pattern typical of their species, and females respond with another pattern which is also species-specific, thereby enticing the male to approach and mate. The female *Photuris*, however, is a predator, and when she detects the flashes made by the male of any one of three other species she responds with the appropriate female version. The unfortunate male flies in to mate and meets a sticky end instead. Not

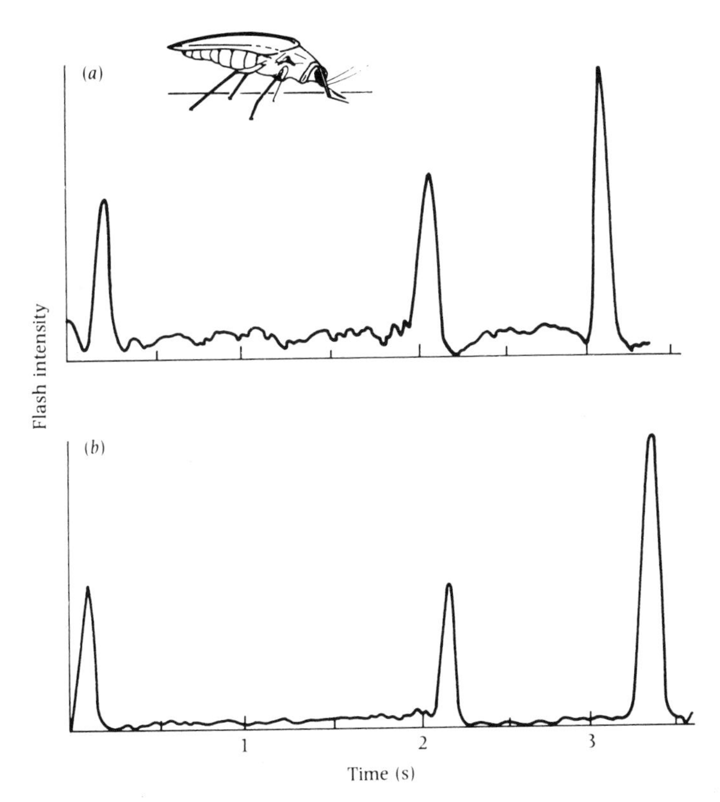

Figure 8.6. (*a*) The male *Photinus* firefly produces two flashes and a female of the same species replies with a single more intense flash. In (*b*) the two flashes are a human imitation of the pattern and the reply is that of a female *Photuris* which attracts male *Photinus* in this way and eats them. (Redrawn from Lloyd, J.E., 1981. *Scient. Amer.* **245**, July, 111–17.)

surprisingly the males of these species have evolved to be somewhat chary in approaching prospective mates!

These examples illustrate the fact that predators and prey may either fit in with their backgrounds and so avoid communicating or may actually communicate with each other, though in this case the information they provide is usually false. That communication does not always involve information that is useful to the animal receiving it is not true just of predators and prey, however. The communication between members of the same species can be interpreted in similar terms.

8.2 Information and manipulation

In the past it was common to refer to communication between animals of the same species as involving the sharing of information for their mutual benefit. Indeed, a lot of communication may be just this. For example, when potential mates meet it is to both their advantages that they communicate to each other accurately information about the species to which they belong, for if they are not members of the same one both of them will waste their breeding effort. Strictly speaking, however, animals will produce signals when this is to their advantage regardless of whether others benefit from receiving them. The examples in the last section showed this well, for prey animals send signals to predators when they gain by doing so and the gain they make is usually at the predator's expense.

Exactly the same rule applies within a species, as we might expect from the idea that animals are essentially selfish, behaving so as to maximise their own inclusive fitness. While the information that passes from one to another when a signal is transmitted may be useful to the one that receives it, this will not necessarily be the case. There are some good examples where it is better to think of animals as manipulating each other rather than sharing information. One of these is in the mating behaviour of the bluegill sunfish (Figure 8.7). Most males of this species do not mature until they are seven or eight years old, when they set up territories to which they attract females. The females are smaller and look quite different. Sometimes two of them may spawn simultaneously with one of these males, sperm and eggs being shed into the water as they swim around together on his territory. Deceit enters into this scheme of things because some males, around 20%, mature at only 2–3 years old, when much smaller than usual. These look just like females and join in the mating of spawning pairs. The territorial male does not drive such a rival off as he cannot tell that he is not a second mate, but actually he is a male and he shares in the fertilisation of the eggs that the female produces. Of course all males cannot mature early, or there would be no territorial ones to attract females in the first place, so this is an example of a mixed evolutionarily stable strategy with a balance between the two possible modes of behaviour. In this example males mimic females and the other males, unable to see through the deception, are manipulated by the mimics. The case of the isopod *Paracerceis*, discussed in Chapter 7, in which the middle-sized morph is a female mimic, can be put in very similar terms. So too may the pied flycatcher example in the last chapter, where the second female to pair with a male may be unable to detect that he is already mated, though there is some doubt about this.

Deception is thought to be possible in cases such as these because there are

Figure 8.7. Three bluegill sunfishes spawning. The large animal at the back is a full-grown adult male. Although they look similar, only the nearer of the two small individuals is a female. That in the middle is a young male which is a female mimic. By this deception these fish avoid being chased away by the large territorial males and are able to steal fertilisations from them. (Redrawn from Gross, M., 1982. *Z. Tierpsychol.* **60**, 1–26.)

no cues that a signal receiver can use to discriminate between the classes of individuals involved. In other cases, however, deceit is impossible because there are ways in which receivers of the signal can see through it. Some good examples here come from cases where it pays animals to appear to be larger than they really are. This is often so in aggressive displays for, the larger an animal appears to be, the more a rival is likely to be intimidated into retreating without a scrap. It is probably for this reason that Siamese fighting fish extend all their fins as far as possible when swimming alongside an opponent and raise their gill covers when facing one (Figure 8.8). Both actions make them as large and striking as they could be. If only some fish displayed like this, as was probably originally the case, they would be likely to have deceived others of the same size into retreat. Thus appearing bigger would have been to their advantage and natural selection would have made this spread through the population until, eventually, all fighting fish looked as large as they possibly could when displaying. At this stage, however, the trick breaks down for each just ends up looking an exaggerated version of its own size. If every fish can increase its apparent size by 20%, all will look that much bigger, but their sizes relative to each other will remain the same. Their opponents will gain accurate information about their size rather than being misled in any way. This outcome has been labelled with the rather curious name of 'superhonesty'.

Figure 8.8. The male Siamese fighting fish is a fine example of an animal that looks larger when displaying (*c* and *d*) than it does normally (*a* and *b*). The displaying fish raises all its fins when broadside to its rival (*c*) and its gill covers when facing it (*d*), thus making itself appear as large as possible. (Redrawn from Simpson, M.J.A., 1968. *Anim. Behav. Monogr.* **1**, 1–73.)

Another nice example where deceit appears to be impossible is in the calling of male toads studied by Nick Davies and Tim Halliday (Figure 8.9). Male toads cling onto ripe females in a posture called amplexus which ensures that they fertilise the eggs as they emerge. Males often fight to take up this position and the male already in amplexus kicks and croaks to repel boarders. Larger males generally win in such encounters, as one might expect from their greater strength. Being larger, however, they are also able to produce deeper sounds, and Davies and Halliday showed that these too had an effect. They silenced males in amplexus by fitting elastic bands round their arms and through their mouths. They then saw whether other males attempted to dis-

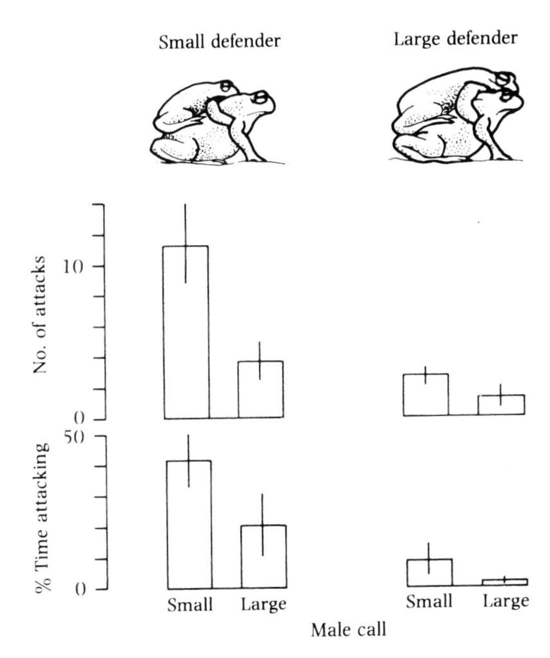

Figure 8.9. Male common toads try to displace each other from the amplexus position on the backs of females. Medium sized males attack more often and for longer when the defender is small than when it is large (the four bars on the left are larger than those on the right). In these experiments the defenders were silenced by passing elastic bands through their mouths and around their forelimbs, and other male calls were played from loudspeakers. No matter what the size of the defender, the attacker would be more persistent if the call played was that of a small toad than if it was that of a large one (the left bar of each pair is larger than the right one). (Redrawn from Davies, N.B., & Halliday, T.R., 1978. *Nature* **274**, 683–5.)

place them. As expected, medium sized males spent more time attempting to displace small males than large ones. However, if the deep croaks of a large male were played from a loudspeaker, a small male in amplexus was harassed much less by his larger rival. Clearly toads can use the calls to assess the size of others and will be less likely to fight with those whose calls are deeper. Here again the information transmitted is accurate and the animal receiving it acts on it appropriately. But why do small toads not cheat, developing deeper voices and so appearing larger than they really are? The answer is probably that the larger an animal is the deeper the sound it is capable of producing. So, just as with the size of fighting fish, if all toads call as deeply as they can, the larger ones will still call more deeply than the smaller ones, and depth of croak will be an accurate reflection of size.

To summarise, communication involves the transmission of information, in the form of a signal, from one individual to another. As we have discovered, this 'information' may be true or false, the important point being that producing the signal should be of benefit to the animal that does so. Sometimes deception may be possible, and the signaller may gain from it, but this is only so if the animal receiving the signal cannot detect that the signal is misleading. Deception tends therefore to be rather rare, for the more animals practice it, the more the recipient will be best to assume the information it is receiving is false. If almost every bird which appears to have a broken wing is really feigning injury, then a fox discovering one would gain most by not following the adult but searching the area for a nest or chicks. Foxes do indeed sometimes do this, suggesting that their prey may be feigning injury just a bit more than is safe!

8.3 Messages and their meanings

A useful way of analysing communication was devised by W. John Smith. He pointed out that it could be looked at either from the point of view of the animal that sent the signal, in which case one would ask what the *message* encoded in it was, or from the viewpoint of recipients, to see what its *meaning* was for them.

Let us return to our singing bird. What is the message incorporated in the song and what meanings may it have for those that are listening? The most obvious point, as every bird-watcher knows, is that song is usually clearly distinct between species, so that it includes information on that to which the singer belongs. Furthermore, song tends only to be produced by male birds in breeding condition, they usually only sing when they have a territory and in some cases they stop singing as soon as they obtain a mate. In such a species the basic message of song may be, 'I am an unmated adult male sedge warbler on my territory and in breeding condition'. In addition, there may be many subtleties incorporated in the message. Signals often vary from place to place, forming dialects like those in human language. Individual animals may also have idiosyncrasies in the form of the signal that they employ. They may have a range of different signals which convey subtly different information. The singing bird may thus also be saying where he comes from, exactly who he is and perhaps even something about his motivation if the songs differ according to whether they are directed at rivals or at potential mates.

Many animal signals are, like song, primarily concerned with reproduction, attracting mates and repelling rivals being their main role in communication. However, especially in species that live in social groups, there are a lot of other messages that it may be useful to convey. Animals that move around in groups often have obvious patterns on them which can be seen from some distance off and enable them to maintain contact with their companions. Call notes can serve a similar function especially where visibility is poor. These are simple sounds and patterns, and they tend not to vary very much. Their message may simply be 'I am here', though usually they also communicate the signaller's species, and there are cases where, despite their simplicity, they vary enough to indicate individual identity. A case in point are the signature whistles of bottlenose dolphins: these are individually distinctive calls, produced by both the group and the lone individual when one animal becomes separated from its group, and which seem therefore to aid in group cohesion.

The alarm calls referred to earlier are an example of another type of message, individuals in this case warning their companions of danger. The 'seeep' call of small birds is most reliably produced by the appearance of a hawk but, in the breeding season, birds with young may also call in this way when alarmed by a dog or a human near to their nest. Most animal signals are like this: the message they convey is a generalised one, rather than being precise like a word in our language. A famous exception is the alarm calling of vervet monkeys (Figure 8.10). These animals have several different calls which they produce when alarmed and three of these are specific to particular sorts of predator: a snake call, a leopard call and an eagle call. Here the information in the call is very precise: the calling animal might as well be shouting 'snake', 'leopard' or 'eagle'.

What do its companions do when they hear one of these calls? In other words, what *meaning* does the call have for them? The answer is that this too is very precisely defined. Animals hearing the leopard call rush up into trees, they respond to the eagle call by running down to the ground and hiding in thickets, and the snake call leads them to approach and look down. We cannot tell if they have a mental image of the predator in question, but they certainly behave as if they were aware of the particular threat it poses. Calls like this are sometimes labelled as 'referential', because they refer to something in the outside world, in this case a predator. By contrast, most animal signals are 'affective', conveying more general messages about the internal state of the animal producing them.

To understand messages we have to study the particular situation in which an animal produces a signal, in other words try to discover what causes it to

Figure 8.10. The opposite and appropriate response of vervet monkeys to two different alarm calls. (*a*) That produced in the presence of a leopard leads them to run into trees, while (*b*) the eagle call leads them to run down into the undergrowth.

do so. In the case of meanings it is the recipient of the signal that we must study: how does it respond when the signal is received? This response may vary from one animal to another. In the case of a species-specific call, for example, other members of the flock may approach so that the group keeps together, animals from other flocks may keep their distance, predators may be attracted with the prospect of a meal, and animals of other species are likely to ignore it as irrelevant to their interests. The same signal can thus have many different meanings depending on the listener and the exact context in which it is received. To stress a point made earlier: the critical factor is that, for a signal to be produced, the advantages to the signaller must outweigh the disadvantages. Otherwise, there is no reason why evolution would favour signalling.

There are thus many different messages that animal signals may convey, though it is unusual for these to be very precise. Most often they are affective signals, which provide rather general information about the motivational state of the signaller, though their features can also incorporate more specific information, such as its species or identity. The number of distinct signals that a species uses is not often large. What makes the signalling system flexible and varied is that the same signal may mean different things to different recipients and the same animal may respond to it in different ways depending on the context in which it is received. Thus, with a comparatively small repertoire of rather stereotyped signals, animals can convey a wealth of information from one to another.

8.4 The form of signals

The message is what an animal encodes in a signal it sends, the meaning is what another makes of it. The signal is the physical form in which the transmission from one to the other takes place. Just as the word 'dog' has neither four legs nor a bark, so the signal is not necessarily related in any very clear way to the message that it conveys. However, some types of signals are better adapted to some functions than others and it is illuminating to consider the characteristics that make them so.

Most animal communication uses sights, sounds and smells, though at close quarters the senses of touch and taste may come into play. Occasional species may use a 'sixth sense', like for example the electric sense of some fish: these can detect each other's discharges and so can send messages using changes in the patterning of their electricity production. Such bizarre senses

	Acoustic	Visual	Chemical
Nocturnal use	Good	Poor	Good
Around objects	Good	Poor	Good
Range	Long	Medium	Long
Rate of change	Fast	Fast	Slow
Locatability	Medium	Good	Poor
Energetic cost	High?	Low	Low

Figure 8.11. A comparison of different sensory channels of communication. (Modified from Alcock, J. 1989. *Animal Behavior.* Sinauer)

are rare, however, and so the three main senses are those with which we are best to be concerned. Figure 8.11 summarises some of the differences between them.

8.4.1 Vision

Changes in posture and colour are the main ways that animals communicate through the visual channel. Vision is the most important sense used by predators that hunt by day, for it is the best for localising and homing in on prey. This means that it is not the ideal sense for other animals to use when advertising for mates. Visual signals also do not travel round corners, so are not good for communicating in a dense environment such as a forest. The distance over which they can be transmitted is also limited by the size of the signaller: small animals cannot be seen a long way off. For all these reasons, the main use of visual signals is in private, short-range communication such as that between mates or between rivals disputing at a territory boundary. When not in use they tend to be hidden so that the animal does not attract predators: most of the bright colours on birds are on their undersides, butterflies fold their wings so that their most colourful surfaces are hidden, and many fish have the capacity to change their colour so that they are less conspicuous. The jet-black ten-spined stickleback chasing off an intruder a few minutes ago is hardly recognisable as the same animal as the small pale grey fish which is now hiding in the weeds from a passing pike. It is interesting that the most brightly coloured animals are found amongst the insects, fish and birds. These, through their capacity to fly or swim, can move in three dimensions and so have extra scope for escaping from predators.

8.4.2 Audition

Sound is not as private a channel of communication as is sight, for noises travel out from a source in all directions and cannot easily be limited so that they are only transmitted to one individual. Sounds can also pass round corners and they move rapidly through the environment, (though admittedly less so than visual signals travelling at the speed of light). Using sound is thus a good way of advertising, provided the animal producing the signal is large enough to generate sufficient noise.

In addition, sound communication has the merit shared with visual signals that a lot of information can be transmitted very rapidly. The pattern of frequency in time which codes this information can be changed with great speed so that one signal can follow quick on the heels of another. Combined with its capacity to travel round corners and over longer distances than visual signals, this makes sound communication ideal for other signalling, such as our language, where a great deal of information is required to be transmitted very rapidly.

Communication by sound provides some nice examples where the exact form of the signal appears to be related to its function. Producing an alarm call in a form the predator cannot easily localise is a case we considered earlier. By contrast with this, advertising animals often produce broad frequency band sounds which are easy to locate and towards which others can therefore orientate without difficulty. However, some sounds travel better than others through different environments, and the habitat in which a species lives may also influence the form of the signal it is best to produce. In woods, for instance, sound echoes off trees so that the pattern in time of a sequence of notes becomes distorted. By contrast pure tones can pass relatively uninfluenced through such an environment. As one might expect from this, bird species living in the tropical forests of central America avoid rapid trills and use tones in their songs more than do the open country species living nearby.

8.4.3 Olfaction

At first sight the production of olfactory signals, or pheromones as they are usually called, might seem to have very few advantages. Smells diffuse only slowly through the environment, their speed and direction of travel being highly wind dependent, and they can carry very little information, for after one smell is released time must elapse for it to disperse before another signal can be employed. However, there are situations where pheromones are ideal. A small animal such as a moth could not be seen or heard from more than 100 metres or so away, no matter how brightly coloured or noisy it was. Yet the male of some moth species can detect the pheromone produced by the female several kilometres away. The chemical involved is a small molecule so that it diffuses rapidly, yet large enough that its structure can be species-specific; the male need only sense a few molecules to start moving upwind to where the female waits for him.

Another merit of pheromones lies in their very persistence: they can continue to signal even when the animal is not there. Many mammals have different scent glands which they use for various purposes. Territorial species often mark the boundaries of their range as a signal to others that the area is occupied. They may do this with special glands, or with scents in their urine or faeces, and a wealth of information can be contained in such signals. The potential intruder may glean, for example, not only where the territorial boundary is, but which individual is occupying it, what reproductive state he is in and even, by sensing how fresh the mark is, how long ago it was that the owner of the territory last went by.

Thus, though olfaction might seem to us the poor sister of the senses as far as communication is concerned, the very persistence and slowness of dissipation of smells make them ideal for some uses.

8.5 Advertising displays

Some of the most striking signals in the animal kingdom are advertising displays, usually those of males advertising for a mate. The lavishness of these has arisen through the process of sexual selection, originally described by Charles Darwin, but later amplified by Ronald Fisher. If females, for whatever reason, prefer males with a particular characteristic, say a larger crest, then the advantage in mate attraction will go to the male whose crest is the biggest and best. Indeed, if crest size is to any extent inherited, then it will certainly pay females to follow the fashion, as mating with a large crested male will give them sons who also tend to have this attractive trait. They will thus have more descendants than females that do not do so. The crest need have no other advantage in itself, provided that females find it attractive. But sexual selection may also favour displays that indicate that their possessors have good genes. For example, females may prefer older males which have proved their ability to survive, males which sing more and which must therefore be good foragers to afford the time, or indeed males with features like large crests that clearly took a lot of energy and materials to build. As time goes on I find the first of these ideas increasingly attractive!

Many of the classic examples of sexual selection come from birds, and in a number of cases it has now been shown that females do genuinely prefer males in which a particular trait is most developed. Many songbirds have repertoires of different songs and in some cases this may extend to hundreds or even thousands of phrases. The sedge warbler, studied by Clive Catchpole, is a good example. In this case a male has a variety of elements from which he composes an almost infinite number of songs. Experiments in the laboratory have shown that females are more likely to show soliciting displays when played tapes of more varied songs than when played simpler ones. Furthermore, in the field, males with larger repertoires of elements were found to attract mates earlier than those with smaller ones.

The classic sexually selected display, discussed by Darwin himself, must surely be the train of the peacock, with its wonderful array of eyespots. It was not until a century after Darwin died, however, that it was shown that it did indeed act in this way. Peacocks mate on leks, where several males display in the same small area, making it easy for females who come to mate to compare between them. As is common on such display grounds, some males get many matings and others very few. Marion Petrie and her colleagues compared between the males in their study group and found that their mating success correlated with the number of eyespots in their trains (Figure 8.12).

Once sexual selection gets hold of a trait, what sets a limit? One important factor is that all such features are expensive. Birds with large song repertoires must devote more brain space to song learning. Birds with large crests or tails or trains must spend energy on building these. Furthermore they have to carry them around, which may be no small feat! Only the bird that has been able to afford to build a lavish ornament, and has not been unduly encumbered by it when chased by predators, will be there, alive and impressive, on the displaying grounds when the females come to visit. The idea that females are most likely to base their choice of mate on costly displays was called the 'handicap principle' by Amotz Zahavi, when he originally described it. He suggested that it was as if males were advertising by saying: 'Look at me, I must be an extraordinarily fit individual to have survived despite carrying this ridiculous handicap around with me!'

The peacock's train is certainly like this, as it serves no other function except that of display and its size depends on the male's ability to obtain the resources needed to build it. On the other hand some displays, of which the

Figure 8.12. The peacock's train: a classic case of a sexually selected signal. (From *Behaviour and Evolution*, ed. P.J.B. Slater & T.R. Halliday. Cambridge University Press, 1994. © Priscilla Barrett.)

depth of a toad's croak would be a good example, are not so much handicaps as genuine readouts of the animal's capacity. Large toads fight better and only large toads can produce deep croaks. So, not all advertising signals have been subject to runaway inflation. But what they do have in common is that they can only be produced by those who can afford them, so that there is no room for deceit.

Another suggestion, which has led to many recent studies, is that females should choose males that are more symmetrical. The idea here is that symmetry of growth is not easy to achieve and can be knocked off course by numerous stresses during development. The individual that is more symmetrical, having successfully withstood those stresses, is likely to be of higher quality than one that has developed with an asymmetry. Stressed individuals are said to suffer from 'fluctuating asymmetry' meaning, not that it changes from time to time, but that the bias goes in one direction in some and the other in others. A good deal of evidence has amassed for such an effect on attractiveness, ranging from the length of tail streamers in swallows to the symmetry of human faces, but it remains controversial.

All this leaves one with the impression of unfortunate males weighed down by the costs of their displays. Is there anything they can do to minimise these? One neat idea, already mentioned in Chapter 3, is that they may benefit from adopting displays that match pre-existing preferences of females, so-called sensory exploitation. A good example here is of the water mite *Neumania papillator*, studied by Heather Proctor. When these animals hunt, they respond to vibrations produced by their prey. Courting males simulate these by trembling their legs and so attract females to them, thus exploiting their predisposition to respond to such stimuli. Another likely example is in bird song repertoires, where males often sing in such a way that successive songs are as different as possible, thereby minimising the chances that females will habituate to them.

As we discussed in the last chapter, there are good reasons why females of many species should be very selective about who they mate with, while males should attempt to mate with as many females as possible. This interplay between choosy females and males doing their best to impress has led to some of the most striking and elaborate displays.

8.6 Animal communication as language

Most animal communication is very different from human language, as the messages it transmits are not at all precise and word-like. They are affective signals, much more akin to those we communicate to each other with our faces: the yawns, smiles, laughs and frowns which indicate to other people what our feelings are. Language is different from this in many ways. But the difference is not so great as might at first sight appear, for some animals are capable of transmitting detailed and accurate information just as we are. It is interesting that one of the best examples is a humble insect rather than an animal more closely related to ourselves. The dance of the honey bee, first described by Karl von Frisch, and for long questioned because it seemed so extraordinary, is certainly one of the most remarkable forms of communication to have been discovered (Figure 8.13).

On a sunny summer's day worker bees fly out from their hive in search of the pollen and nectar on which the colony feeds. If a worker finds a good food source within about 100 metres or so of the hive she will return and perform a *round dance* on the vertical surface of the honey comb. Her movements attract other workers and they crowd round her sensing the smell of the flower on which she has been feeding. They will then fly out and search for this smell in the neighbourhood of the hive, thus homing in on the food supply that she has found.

Figure 8.13. The waggle dance of the honey bee gives other workers information about the location of food. The nearer the source the more rapidly the dance is performed. The dance is in the form of a figure of eight and the orientation of the central run of this to the vertical in the hive is the same as the angle between the direction of the sun and that of the food outside. The forager shown here is dancing directly up the hive, thus indicating a food source in the same direction as the sun. (After von Frisch, K., 1966. *The Dancing Bees*, 2nd edn. Methuen, London.)

The round dance does little more than stimulate other bees to take an interest in the dancer, but the information in the *waggle dance*, which workers returning from more distant sites perform, is much more detailed. In this dance the bee performs a figure of eight, looping over the surface of the comb, first in one direction then in the other, and waggling her abdomen as she goes. What von Frisch discovered was that this pattern informs other bees of both the direction and distance of the source of food that the forager has discovered. Direction is indicated by the orientation of the dance on the surface of the comb. If the food is directly towards the sun, the bee will make the middle run of its figure of eight vertically up the hive, and if in the opposite direction it will go straight down. In other words the angle between the sun and the source of food as the bee emerges from the hive is the same as that between vertical and the direction in which it dances when back inside the hive. A remarkable feature of this system is that workers need neither see the dancer nor the sun to be recruited. It is dark inside the hive so they can only sense the direction the dancer takes by touch. Outside the hive the sun may be hidden behind cloud but the bees will still take up the correct direction as long as a small patch of blue sky is visible. This is because they are sensitive to the pattern of polarised light in the sky and can tell where the sun is from this.

So much for direction. It is the waggles that the forager makes as it dances that indicate distance. A worker that has had to fly a long way dances rather lethargically, whereas one that has found food close to dances with vigour and so produces more waggles per unit time. Each waggle is accompanied by a burst of sound and it is probably the rate of these sound pulses that is used by other bees to assess distance.

The bee dance has several notable attributes, two of which are features one might otherwise think of as unique to human language. One is that it is *symbolic*: information about distance and direction are encoded in features of the dance in a stylised way. The other point is that the bees are communicating, as we often do, about events which are distant in time and space from where the communication is taking place. In other words, in the darkness of the hive they are telling each other where the food is even though this may be some kilometres away.

The communication of bees is remarkable, but comparing it with human language is a somewhat sterile pursuit because the two systems are obviously vastly different in complexity. An even more contentious subject has been the comparison between our language and that which people have succeeded in teaching to chimpanzees and gorillas. Initial efforts to 'teach chimps to speak' were to no avail, but probably partly because they lack the vocal apparatus necessary to produce the sounds we can master. More success has been achieved with methods which depend on the skills they have with their hands. In the best known case Allen and Beatrice Gardner taught a young female chimp called Washoe nearly 200 of the gestures used in AMESLAN, the language of the deaf in America (Figure 8.14). She used them in appropriate contexts, strung them together in novel ways, could grasp concepts such as 'open' (applied to both a door and a matchbox) and 'dog' (applied to things as different as seeing a chihuahua and hearing a bark).

Subsequent studies of this type have used a variety of different techniques on various species of animal. The work has extended to bottlenose dolphins, sea lions and grey parrots (which have the advantage of being able to 'speak'!), as well as gorillas and bonobos (the other species of chimp). They have used

Figure 8.14. A chimpanzee using American sign language (AMESLAN). The chimp shown here is Moja, one of Washoe's successors. She is being questioned about a birch tree and has put her arms in the correct configuration to signal the word 'tree' (photograph by courtesy of B.T. and R.A. Gardner).

computer keyboards or symbols to communicate with their human carers. However, the work has tended to move towards studies of comprehension rather than production, on the view that this is where the essence of language lies, and they have also responded to criticism by being very careful about controls. These studies have revealed some remarkable capacities. For example, the bonobo Kanzi, studied by Sue Savage-Rumbaugh and her colleagues, has proved capable of following a detailed series of instructions (given by someone wearing a space helmet to ensure that no glance at an object or other facial cue could be involved). Kanzi was not formally trained, but kept with his foster mother while efforts were being made, unsucessfully, to train her, and picked up both keyboard use and understanding spoken instructions on the side. Bonobos seem particularly good at this and, as with so much else in development, starting young is an important ingredient of success.

Whether any of the feats of animals amount to language in the human sense has caused a great deal of heated argument which has shed next to no light on the issues. It will always be possible to point to ways in which such animals fail to match up to human achievements for the simple reason that they are not human. But perhaps two things are worth pointing out. First, whether one wants to label them as language or not, the communicative skills that animals such as Washoe and Kanzi have achieved are obviously remarkable. Second, it is very unlikely that animals possess such capacities and yet leave them lying latent. In the wild chimps are sociable animals, living in rather loose groups. The sounds, gestures and facial expressions that they make may not appear complicated to us, but there may be a hidden wealth of meaning in them which will only be unravelled when we decide to invest as much in learning to understand their language as we have already done in trying to teach them to speak ours.

8.7 Conclusion

Animal communication is one of the liveliest areas of research into behaviour at present, partly because it bears on the relationships between animals and the extent to which they are able to influence each other to their own ends. There is no doubt that some very sophisticated things are going on here, and the theory underlying animal communication has been in continual ferment for quite some time. A lot of the natural history is pretty fascinating too! In moving on to social organisation, as we do in the next chapter, we will not leave the topic totally behind, for communication is of course the stuff of which the interactions between animals in social groups are made.

9

Social organisation

Strictly speaking, it is questionable whether one should refer to animal societies as 'organised' at all. As earlier chapters have emphasised, natural selection acts on individuals, favouring the genes of the better adapted, but not selecting for one group rather than another. Societies have thus not evolved through the action of natural selection upon them, but they have emerged as a result of the way selection has affected the behaviour of their individual members. Nevertheless, different animals occur in very different social structures and, though each may be selected to maximise its own inclusive fitness, elaborate societies may result from the way in which such individuals interact with each other.

In this last chapter we will consider some of the features of these societies, how they differ from each other, how they come to have a semblance of being organised and what social life involves for animals that belong to such groups. But first we must consider why animals would want to live in groups at all, rather than as single individuals on a territory or home range of their own.

9.1 Territories and home ranges

Some animals, such as cats and other solitary hunters, spend most of their time on their own, though even they must come together in pairs for reproduction. Nevertheless, the presence of others would hamper their mode of food finding and their life is thus essentially a lone one, wandering widely over a range on which they hunt. Sometimes such 'home ranges' overlap so that different individuals may use the same area and, if those of males habitually

Figure 9.1. The mosaic of territories occupied by mudskippers (*Boleophthalmus boddarti*) at low tide. These fish construct mud walls between their territories, so that their shape can be seen. The walls function to reduce visual contact and so aggression between neighbours (photograph by D.A. Clayton).

overlap with those of females, they may meet for mating without either leaving its patch. Sometimes, however, animals live as individuals, pairs or larger groups on 'exclusive territories' from which they repel other members of their species, at least those of the same sex as themselves. The boundaries between such areas can be fiercely contested and often end up being quite sharp (Figure 9.1). A nice example here, where the borderline can actually be seen, is in the mudskipper *Boleophthalmus*, a small fish which has the capacity to live out of water for periods when the tide is out. These fish are highly territorial and build mud walls between the burrows they occupy so that the mud is a mosaic of areas, each fenced off like a suburban garden. When the tide is out the walls preserve some water within the territory and this may help in their respiration and in the growth of the microscopic plants on which they feed. But another function of the territory seems to be as a display area, for the males leap in the air and this attracts females to mate with them.

There are several possible reasons why animals might defend territories. The two most common ones are that it enables them to court and mate without interference from others and that it allows them to sequester a food supply which they can use systematically without others raiding it. The males of many small songbirds set up such territories, and these are usually large enough to provide sufficient food both for themselves, and for their mate and offspring.

Defending a territory as a source of food is most practicable if the food supply is rather constant and evenly spread. If this is not so, then the area required to feed the animal, or a family if it is during the breeding season, may be so great that it is just impossible for one animal to patrol its boundaries and ensure that others are kept out. The distribution of food and the amount of it that an individual can defend can have a profound influence on social structure. For example, it has been argued that polygyny, a single male having more than one female, may have evolved in territorial birds where some territories are very much richer in food than are others. As a result a female may breed more successfully by being the second one of a male on a rich territory rather than the sole partner of a male whose territory is poor. If food is patchy in time and space this may also have implications for social structure. European badgers live in groups of 2 to 12, on ranges that also vary enormously in size. Their food is largely earthworms, and the abundance of these varies from place to place and also with time in the same place. The range size in a particular area is such as to ensure that there is always a productive patch somewhere within it, and the group size is related to how good a supply of worms there is overall within the range.

It is probably largely such considerations of food availability that determine whether animals maintain territories or live on rather looser home ranges. But, whichever is the case, the numbers of individuals that occupy the territory or range may vary from one up to a sizeable group, all moving around together and sharing the resources of the area, or all foraging more or less independently of each other. Why should it benefit selfish individuals to live in such groups?

9.2 Why live in groups?

Most of the reasons that have been suggested for group living relate to ways in which grouping helps in the finding of food and in defence against predators, and we will discuss these two ideas in turn.

9.2.1 Defence against predators

Group living yields a number of advantages related to defence. While a group of animals is bigger than a single one, and so can be spotted from further off,

Figure 9.2. Within a flock of geese a few individuals have their heads up, scanning for danger, at any one time. Those that are foraging can thus do so more safely as they will be warned should a predator approach. The larger the flock, the less time each individual needs to spend with its head up (photograph © M. Owen, Wildfowl Trust, Slimbridge, UK)

if the predator only eats one prey at a time it will on average take longer to find a meal if the prey are clumped than if they are evenly spread out. Furthermore, there are more pairs of eyes in a group so the predator is less likely to creep up undetected. To some extent this is offset by the fact that animals in groups are able to be less vigilant than solitary individuals. Lone geese, for instance, raise their heads much more frequently while cropping the grass than they do when feeding in groups. Because there are more pairs of eyes perhaps they can afford to keep their heads down and feed for longer periods with the security that others will raise a warning if danger threatens. Equally plausibly, they may show less vigilance simply because there is less risk to each animal in a group than when it is on its own (Figure 9.2). Both of these factors probably contribute. In minnows, for example, when a shoal is approached by a model pike, the proportion feeding declines sooner in larger groups suggesting that they detect the model earlier. After experience of attack by a pike, they form bigger shoals and these are maintained even 24

Figure 9.3. To illustrate the idea of the selfish herd, Hamilton used the example of frogs sitting on the edge of a pond from which a snake sometimes pounced and took the one nearest to it. Each frog thus has a zone of danger that stretches half way to its nearest neighbour on either side. An individual can minimise its own threat by moving so that others are close to it on both sides, as the dark animal is shown to do here. If all of them do this, they will end up in a tight group. (After Hamilton, W.D., 1971. *J. theor. Biol.* **31**, 295–311.)

hours later. When a pike appears minnows make so-called predator inspection visits, swimming rapidly up to it and away again, and they will go closer to it when they have a shoal to return to than when alone. These points suggest that they are both more vigilant and have less to fear when in a group.

The lower risk to grouped animals was pointed out by William Hamilton in a paper called 'Geometry for the selfish herd'. He made the simple point that, if predators just take one prey at a time, the best prey strategy is to keep another individual between oneself and the predator (Figure 9.3). Prey animals that do this will tend to form groups simply because being in a group minimises the area of danger to each animal. Being in a group can also confuse a predator because it has difficulty fixating upon and chasing a single individual. Thus a pike put in a tank with a single stickleback will rapidly catch it but, if there is a shoal of sticklebacks, catching the first one takes much longer even though the sticklebacks do not combine to defend themselves actively in any way. Some animals do cooperate in defence, however, giving yet another advantage to group living. Musk oxen set upon by a pack of wolves will form a circle facing outwards so that the wolves are confronted by a solid wall of horns and cannot reach any of their vulnerable flanks.

A final example of grouping geared to predator defence is a very curious one and this is in the nesting behaviour of ostriches. Several female ostriches lay in one nest, but only the first to lay incubates: she accumulates a much larger number of eggs in this way than she could lay herself. After the eggs hatch the female may have around 25 chicks following her as she sets off across the savannah, but later she and her mate may collect even more as a result of chasing other hens away (Figure 9.4). Brian Bertram, who studied this extraordinary system, even recorded one group in which two adults were accompanied by 105 chicks.

Figure 9.4. An ostrich surrounded by a group of chicks. Some of these will have hatched from eggs the female laid herself, some from eggs other females laid in her nest, and some of them may have hatched in other nests and joined the group later. The advantage to adults in tending these creches is thought to result from a dilution effect: if a predator takes one chick, the more there are in the group the less likely is that that chick to be one of the adult's own (photograph © F. W. Lane).

What advantage can there be to a female ostrich in caring for the eggs and chicks of others? The answer seems to be that chicks survive better in larger groups so that her own chicks benefit by a 'dilution effect'. Predators such as jackals will eat ostrich chicks, but the more of these a female has with her the less likely is the one taken to be one of her own. This cannot be the whole story, however, otherwise a female would gain whether she cared for her young herself or let another bird do so. At the stage of incubation it seems that there *are* advantages to the hen that sits, for she can recognise her own eggs and she gives preference to them so that they remain in the centre of the nest and are more likely to hatch. She is not therefore being a generous nanny helping out others, but is running the crèche purely for her own advantage. It remains to be seen whether females with chicks also gain from giving care themselves rather than leaving them to be looked after by others.

Figure 9.5. Grouping as an aid to feeding. By hunting in groups small carnivores, like the wild dogs shown here, are able to tackle prey much larger than possible for a single animal (photograph by P.A. Jewell).

9.2.2 Finding food

A further range of benefits to group living is related to feeding. Minnows that can see others in an adjacent tank do not necessarily gather at the end of their own tank nearest to them, but they are much more likely to do so if the neighbours are feeding. Thus, if food is clumped, as it very often is, animals that feed on it tend to become so too. Seeing another individual feeding may be a good indication of where food is abundant so groups may form for this reason alone without there being any advantage in being a member of a group once there. Indeed the animals may interfere with each other's feeding efforts but, if they are conspicuous, there is no way in which a feeding one can stop newcomers arriving to share in the spoils.

Not all feeding groups are passive clumps, however. While many predators, such as frogs, snakes and domestic cats, are solitary hunters, using stealth to catch small prey, some others, like lions, hyaenas and hunting dogs, cooperate in the hunt (Figure 9.5). This may increase the chances of success, because individuals can take turns in chasing and because they can spread out and so limit escape routes. Golden jackals hunting Thomson's gazelles are four times as successful in pairs than when on their own: of course there are twice as many mouths to feed, but it still means that each will get double the food. Hunting in groups may also enable animals to tackle far larger prey than could an individual on its own: a pack of hyaenas can pull down a zebra or a wildebeest, or some of them can distract a rhino mother while others attack her calf. In African wild dogs, Scott and Nancy Creel found that the success of hunts rises with group size, both in terms of the numbers and size of prey killed. The food intake of each dog is not as high in groups of the most usual size as it would be in larger or smaller ones, but the normal size of group is more economical as far as the amount of running in a chase is concerned. When costs as well as benefits are taken into account, the wild dogs seem to be doing the right thing.

A further advantage of grouping may also be important in animals that are not carnivorous. This is that groups may be able to make use of resources in a more systematic way than can isolated individuals. The bumblebee flying from flower to flower may well be following close behind another one and so getting less reward than it would otherwise do. If individuals form flocks and move round their range together, as is often true of small birds outside the breeding season, they can be sure that they are all arriving in a fresh area where the food has had a chance to replenish itself since last they came that way.

One final advantage that animals may gain from living in groups is that they may glean information from each other. Some bird species, such as starlings, gather in huge roosts each night (Figure 9.6). Being in a large group probably helps to warn them of the approach of nocturnal predators such as cats and foxes. But another benefit that they may gain is in information about where the best sources of food are to be found. It has been suggested that animals that have fed badly one day, by watching others departing from the roost next morning, may join a group that is going back to a place it had found profitable. The theory that roosts may act as information centres is, however, controversial: in particular it is not at all clear why foragers that have found a good supply of food should selflessly return to the roost to share their news with others. But, in more permanent groups than those of roosting birds, where kinship and cooperation may be involved, individuals may certainly benefit from the knowledge of others. In times of drought the most aged member of a monkey troop may have information about water holes that do not dry out that others would be too young to remember.

Many monkey groups occupy quite large territories and move over them in a single party searching for the fruit or buds on which they live. They too may gain from the systematic foraging that group living allows, but their groups are more integrated and the relationships of individuals within them more varied than in a flock of birds or a herd of deer. The advantages of food finding or

Figure 9.6. Many birds, such as the starlings shown here, can roost in huge numbers. It has been suggested that these gatherings may act as 'information centres' so that individuals that failed to feed well on one day can find a better supply on the next by following others as they fly out in the morning (photograph by C.J. Feare).

predator avoidance may explain why they, like many other animals, come to live in groups, but do little to account for the richness of the interactions between the group members. It is to various facets of the behaviour of individuals within groups towards one another that we shall now turn.

9.3 Kinship

Not surprisingly, animals that live in social groups tend to be related to one another. Young ones are born into the group to which their mother and father belong and may often stay in it to become parents themselves. However, if this was always so, animals in the group would become more and more inbred. Too much inbreeding leads to young which are less viable so it is discouraged by natural selection. What usually happens therefore is that young animals of one sex or the other move away from the group in which they were born as maturity approaches and so do not mate with close relatives. In birds it is usually the females that disperse to different areas, whereas in mammals the males are more likely to leave, although there are quite a few exceptions to this rule including species in which both sexes move away.

Birds tend to be territorial and monogamous (give or take a bit of extra-pair mating!) and it may benefit males, as the territorial sex, to stay around and get to know the area well. By contrast, many mammalian species are polygynous and matriarchal, females staying with their mothers while males leave to find other groups which they can join. Lions are a case in point here. As was mentioned in Chapter 7, males leave their natal group and go in search of another that they can take over. Females, however, stay in the group in which they were born so that the lionesses in a pride are usually closely related to one another. They often give birth at around the same time and it has been found that they will suckle each other's cubs, something one would only expect if they had an interest in them through being related. Furthermore, again in line with what kinship would predict, they suckle their sisters' offspring more than those of their cousins.

Alarm calling is another behaviour where kinship is important. Making a noise when a predator is around is obviously a risky thing to do, and one would expect to find some advantage to offset this. Many birds only call when they have young in the nest, so saving these from being discovered seems to be the benefit. Using a stuffed badger as a 'standard predator', John Hoogland has shown that black-tailed prairie dogs produce more alarm calls when they have close relatives in their group than when they do not. Males called a lot in their natal group but ceased to do so after they moved elsewhere only to start calling again once their own young were born in the group that they had joined. All these findings fit in well with the idea that animals are more likely to call when relatives may benefit.

As far as the relatedness of their members is concerned, the most complex groupings are amongst the social insects. In the honey bee, only the queen lays eggs and all other females in her hive are sterile workers. The eggs the queen lays are of two sorts: unfertilised ones which will develop into fertile males (drones) and fertilised ones which usually develop into infertile females (workers) but, if nourished only on royal jelly, will form a new queen. This happens when the old queen dies or when the colony has grown to a point where it must split. A peculiarity of this breeding system is that drones have half the number of genes that queens or workers do and, when they fertilise an egg, *all* their genes pass to their offspring instead of just half of them. As a result of this mode of inheritance, which is known as haplodiploidy, workers share ¾ of their genes with their sisters rather than ½ as most animals do. It is thought to be for this reason that it benefits them to raise sisters that will become queens rather than having daughters of their own. Through a quirk of genetics their sisters are more closely related to them than their daughters would be and are thus a better investment in the future of their genes.

The social system of bees is thus extraordinarily intricate, and this is

Figure 9.7. A colony of naked mole-rats packed tightly together in its burrow. The social system of this species is more like that of a social insect, with castes and adults that do not breed, than those of other mammals (photograph by J.U.M. Jarvis).

probably because these animals are especially closely related to others in the hive to which they belong. However, having their unusual mode of inheritance is not a prerequisite for a social system of this sort. Termites are not haplodiploid yet they have a colony structure which is just as complex. More remarkable still is the naked mole-rat, for this is a mammal which has colonies much more like those of a social insect (Figure 9.7). These animals live communally in groups averaging around 80 individuals which share a burrow. For most of the time they huddle together in one underground chamber, and it is here that the only female to breed has her young. She is large, and males approaching her size may mate with her, but most other large animals in the group neither breed nor forage: their main role seems to be to keep the colony warm. Smaller individuals are like the workers of social insects. They nest build and move around through the burrow system foraging for roots and tubers which they bring back for all the animals to eat. If there is a disturbance, all the members of the colony join in the task of carrying the young off to safety. As well as the food that the workers bring back being shared, the young obtain nourishment from adults by eating their faeces. Individuals also spread their urine round the colony and groom with it. At first this was thought to be just like the social insects, in which queen substance is spread throughout the colony and stops the workers from rearing other queens. It is true that, in mole-rats too, if the breeding female dies or is removed, another one comes into reproductive condition very soon. However, experiments suggest that the 'information' in her urine and faeces is not what suppresses the breeding of others. Instead, they must have direct contact with her. She has a behaviour pattern called 'shoving', incidentally directed more to less closely related animals, which seems to keep them all in their place.

Mole-rats are not very mobile and their burrow systems are isolated from each other. Most of the animals within a colony are relatives and DNA fingerprinting has estimated their coefficient of relatedness (r) as 0.82. Given that the figure for full siblings is 0.5 in an outbred mammal, and for identical twins 1.0, this shows just how close the relatedness is within one of their colonies. This closeness undoubtedly accounts for the common interest that colony members have in the offspring of the breeding female.

Differences in the degree of kinship between animals in a group may therefore have a marked influence on that group's social structure. This may happen without animals being able to recognise their kin in any way, but simply because the chances are high of an individual they meet being a close relative. However, in practice, the recognition of kin does have a marked effect. Whether or not an intruder gets past the guards at the entrance to a sweat bee nest depends very much on relatedness, the chances of getting past correlating almost perfectly with r. In naked mole-rats, while their groups may wallow contentedly in urine and faeces, it is blood they swim in when the tunnels of two groups happen to bump into one another. Some recognition like this may simply depend on experience: there is a good chance that animals reared in the same nest are siblings and they may therefore behave towards each other with appropriate altruism. But this does not account for it all. Young spiny mice that are cross-fostered to other litters do socialise more with their foster siblings than with strangers thereafter. But, animals from different litters but the same mother, which are thus related to each other but have never met, are still more likely to socialise than ones that are unrelated. What is going on here is a process called 'phenotype matching', one animal assessing its relatedness to others by comparing them with its parents, siblings, or indeed itself.

What cues may be used to do this? One thought to be particularly important in mammals is smell. A part of the genome, known as the major histocompatibility complex or MHC, is especially variable and is responsible for cell surface proteins: these vary a lot from one animal to another as part of their defence against infection. But breakdown products of these proteins also appear in the urine, leading to variations in smell which correlate with relatedness. Female mice prefer to mate with males that differ from themselves in the MHC: as well as helping to produce offspring with more varied cell surface proteins, and which are therefore more resistant to infection, this may well be an important mechanism to ensure outbreeding.

Kin recognition is obviously potentially important when it comes to mating, as a means of avoiding inbreeding. But it may also have a role to play in cooperation. As we saw in Section 7.3 the benefits of helping another individual are very much influenced by relatedness. Let us take an extreme example. Amphibian tadpoles often preferentially form shoals with siblings. Some grow large but others fail to do so and are clearly not destined to survive. At this point, if given a choice, large marbled salamander larvae preferentially eat small ones to which they are related. This may seem odd, but perhaps it should be viewed the other way round. If you are going to die, the best way to enhance your inclusive fitness is to feed yourself to a relative!

9.4 Cooperation

Not all cooperation involves kinship. If two jackals help each other to capture a Thomson's gazelle which neither on its own could catch up with, then both of them show an immediate gain and the cooperation is of obvious advantage whether they are related or not. Rather more difficult to explain are cases where unrelated animals are generous without getting anything in return. A likely mechanism here, though one of which there are as yet very few examples, is referred to as 'reciprocal altruism'. This is where one animal helps another member of its group on the expectation that the favour will be repaid at some later date. It is just like cooperation, except that it involves a time lag. It is obviously a very important phenomenon in humans, for many of our relationships are based upon kindnesses which, while they may run in one direction in the short term, balance out in the long run.

For reciprocal altruism to evolve there are two important prerequisities. First, animals must stay together long enough for it to be likely that an opportunity for repayment will arise. Second, they must be able to recognise each other as individuals. If they cannot do so then 'cheaters', animals that receive but refuse to give in return, will remain undetected. As these get all the benefits without bearing any of the costs they will be at an advantage unless they can be excluded from the system. Given these points, it is not surprising that few examples of reciprocal altruism have been described in animal

Figure 9.8. A vampire bat. In this species an individual will sometimes share the blood on which it has fed with another group member. This benefits both of them, as the donor will usually be repaid by the recipient on a later occasion (photograph by U. Schmidt).

groups. Perhaps the best one comes from the study by Jerry Wilkinson on food sharing in vampire bats (Figure 9.8), a species that would not come to mind as an obvious example of an altruist! These animals live in groups and, if an individual fails to find a blood meal one night, another group member will regurgitate blood for it and so save it from starving. Usually the two animals in this arrangement are kin but, where they are not, they are most likely to be ones in which the recipient gave blood to the donor on a previous occasion. Thus, to explain the food sharing of vampire bats, it is necessary to invoke both kinship and reciprocal altruism.

Another probable case is in the breeding of green wood-hoopoes, a species in which there are helpers at the nest but not all of them are relatives. Two factors are important here: helpers tend to remain in the group to become breeders, and breeding success depends critically on having helpers. J. D. and S. H. Ligon suggest that, in helping breeders to rear their young they are ensuring that a supply of helpers will be there when they become breeders themselves. Their altruism is reciprocated by the help those individuals will provide for them.

In reciprocal altruism there is always a risk that the favour will not be returned. But, even where two animals cooperate to their immediate mutual advantage, there is the danger that one will make off with all the spoils and leave the other stranded. This situation can be modelled in the same way as ESSs, using game theory, the relevant game here being called the Prisoner's dilemma. Its critical feature is that two individuals do better if they cooperate than either would on its own, but the greatest payoff is to one that can persuade the other to cooperate but does not do so in return. It thus bears only half the costs but gets all the benefits. This situation has been studied a great deal, particularly using computer simulations to predict how pairs of animals (or people) that meet repeatedly, as they would in social groups, should behave. While the theory is now well ahead of our knowledge of how animals actually do behave, the results suggest that cooperation should evolve in close knit social groups. For example, one of the best strategies identified so far is called Tit-for-Tat: cooperate on the first encounter with an individual, and thereafter do what they did on the previous occasion. If everyone plays that game then they will all continue to cooperate with each other.

9.5 Dominance

It is a rare group of animals in which all is peaceful and harmonious, each assisting the others and food being shared equally around. More often fights and squabbles occur and some animals do rather better than others, getting more of the food and doing more of the mating than their companions. Where one animal habitually beats another in fights, the first is said to be dominant and the second subordinate. In many animal groups one individual, usually a large and strong male, dominates all others. This alpha male has priority of access to any choice food that the group comes across and to females within it that are receptive.

An interesting feature of dominance is how clear cut it often is, the alpha individual always defeating all others, the beta all but the alpha, and so on in a straight hierarchy down to the most lowly individual who wins no fights at all (Figure 9.9). This has been found even in quite large groups of chickens, though the occasional triangle may develop in their peck order, so that A beats B, B beats C and C beats A. If one sets up a new group of animals, they tend to squabble at first, but after a little while they settle down, each of them knowing its place, chasing off those beneath it but unprepared to challenge

black bib, which indicates their status. Subordinates with their bibs dyed black do not get away with it, as they are persecuted by the real dominants, so the badge seems to be a genuine indication of fighting ability. Sievert Rohwer, who has studied these birds, suggests that there are two strategies. The dominants do not search for food but displace subordinates that have found it. The dominants but not the subordinates have the cost of fighting; the subordinates but not the dominants have the cost of searching.

In many species males are bigger, stronger and more aggressive than females and, if both sexes form a single hierarchy, as happens for example in a group of weaver birds, the females tend to be dominated by the males. But it is not always the case that hierarchies are straightforward listings of animals in order of size or strength. Sometimes one animal may beat another on some occasions and be beaten by it on others. This may depend on exactly what they are fighting over, but the circumstances leading to such 'back pecks' are not yet worked out. In other cases the relatedness of animals within a group may affect their status. Rhesus monkeys live in matriarchal societies (Figure 9.11), young females staying with their mothers to breed while their brothers disperse elsewhere. Several generations may occur in the same group and, in this case, the older females are dominant to their daughters and grand-daughters. Within a generation, however, dominance is in inverse order of age, a female's youngest daughter coming to dominate her older sisters as soon as she is mature. She is able to do this partly because her mother manages the situation and exerts a certain amount of favouritism over her more recently born young.

Dominance hierarchies are a feature of animal societies that gives them the impression of being highly organised for, at least in cases where they are clear cut, each individual knows its place, defers to those above and ousts those below. Open aggression is rare once positions are settled and a threat display, or perhaps just a slight movement of intent, may be sufficient for the superior to assert its position. As with other aspects of social behaviour, however, the organisation is more apparent than real. The hierarchy exists because individuals have discovered that they are able to dominate those below them but not those above.

9.6 Synchrony

If animals are to stay in the groups to which they belong they must do roughly the same thing at the same time as each other. It is safe to groom or sit quietly when others are sleeping, but the animal that falls asleep when the group is

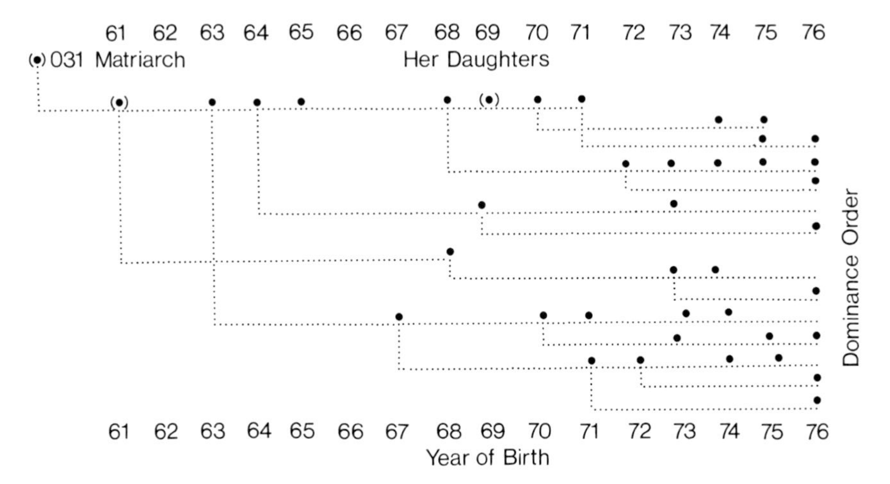

Figure 9.11. Dominance relations in the matrilineal society of rhesus monkeys. Here the diagram shows the relations between the female descendants of one female (031) studied on Cayo Santiago island. For simplicity males are omitted, though many were, of course, born during the period. Brackets show animals already dead at the start of the study. Six of 031's eight daughters had female offspring, and in some cases there were also subsequent generations. The horizontal level these are shown at indicates their dominance status. Note that daughters usually take on the dominance position of their mothers; within a family no daughter overtakes her mother in dominance, but younger daughters and their offspring tend to be dominant to older ones. (Modified from Datta, S., 1988. *Anim. Behav.* **36**, 754–772.)

foraging or starts to explore while the rest are asleep will quickly find itself alone.

Synchrony of behaviour is achieved in a number of ways. One of these is that, quite simply, animals tend to do certain things at a particular time of day even if entirely on their own (Figure 9.12). Most species that are active by day show a burst of moving around and looking for food in the hours immediately after dawn, they will then rest and groom during the mid-day period, and perhaps show a brief resurgence of activity just before dusk. Nocturnal species tend to show the opposite pattern: a peak of activity just after the sun goes down and a more minor one just before dawn.

Patterns like these, on their own, will lead to different animals showing broadly similar behaviour at equivalent times. But the phenomenon of social facilitation sharpens up such timing. If a hungry rat is given food in the presence of a companion that has eaten its fill, the second animal is likely to start eating again, albeit in a rather desultory way. If one bird is given water, others

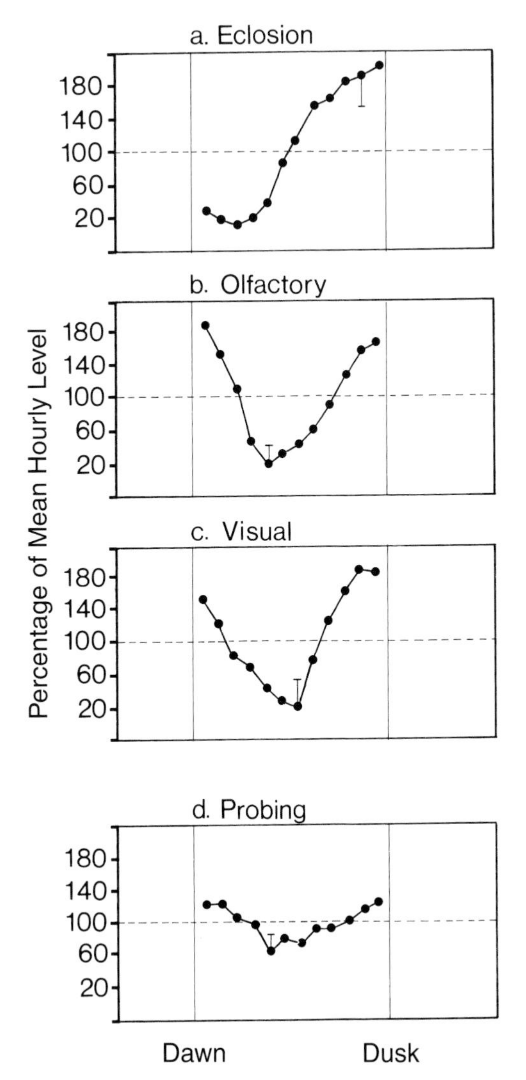

Figure 9.12. The behaviour of animals can change dramatically from one time of day to another. In species that are active by day, various activities tend to peak early in the morning, with a lesser peak as dusk approaches and a period of quiescence in the middle of the day. This is illustrated here by hourly levels of various activities shown by tsetse flies, most of which have a trough in the middle of the day. All animals in a species tend to show similar changes, so the behaviour of different individuals will be synchronised without them necessarily being influenced by one another. (After Brady, J., 1988. *Adv. Study Behav.* **18**, 153–91.)

able to see it bathing will start to wash themselves even though their water has been unchanged for days. And we all know how infectious yawning is! Social facilitation is a potent force leading animals to show the same behaviour as their companions.

It is perhaps curious that leadership is not usually involved in the synchrony of behaviour that animals show. Large and dominant individuals may determine when a group moves and when it rests more than others, but much of the patterning of group behaviour depends more on consensus than on despotism. The goose which takes the lead as they fly off in V-shaped formation is a different bird from one flight to the next. Even in primates, where the benefits of age and experience might suggest that it would be beneficial to follow some individuals more than others, patterns of movement are far from well-ordered. As a troop of yellow baboons moves along, the order of individuals in it is essentially random, though there is some tendency for both adult males and juveniles to be in the front third. In hamadryas baboons, where groups leaving the cliffs on which they sleep can be very large, Hans Kummer has described how different individuals may start off in different directions and then wait to see whether others follow. While the group may not have a single leader, some individuals may exert more influence than others by refusing to follow where they are going.

9.7 In conclusion: animal behaviour and human behaviour

It is appropriate to end with a few words about ourselves. We have occasionally alluded to humans during the book, and many of the animal examples have illustrated abilities that may seem to come close to our own. Indeed some quite simple animals can carry out what look like remarkable feats. One example is the young chaffinch that memorises a complex pattern of sounds one summer, remains silent through the winter, and then produces a perfect copy the following spring. Perhaps more amazing is the ability of some food storing birds and mammals to hide literally thousands of food items and to recall where they have put them, returning to these specific places later on. These are wonderful abilities fitting the animals in to their way of life, and in these particular areas they are wizards. But try them on a different sort of task, for which natural selection has not equipped them, and they will fail abysmally. This is sometimes referred to as indicating 'constraints on learning', as if a general learning ability was to be expected, but some deficits were found. It is better to think of them as having 'special abilities' to stress the fact that selection has equipped them with these specific skills. This is a far cry from the more generalised intelligence which we can bring to bear on a great variety of tasks, from chess to motor cycle maintenance, which certainly did not exist in the environment in which we evolved.

The difference between humans and animals may be largely one of degree rather than of kind, but it is there all the same. Our language puts us in a very different game from the rest of the animal kingdom, with its immense potential for cooperation and cultural transmission. While it may be entertaining to look at our behaviour as if it was moulded only by natural selection, as some sociobiologists and evolutionary psychologists are prone to do, there are dangers in this 'nothing-but-ism'. There is much more to human behaviour than that! If there is one message of this book that I would like to have got across more than any other it is that, as far as behaviour is concerned, genes influence everything but determine nothing.

Suggested further reading

Two recent books taking the subject further than possible here, both of them excellent, are those by Manning & Dawkins (1998) and, at a more advanced level, by Alcock (1998). Dawkins (1995) does an excellent job of discussing particular knotty issues and includes chapters of relevance throughout this book. For a more historical perspective, the book edited by Houck & Drickamer (1996) reprints a variety of the papers that were influential during the first 100 years of the scientific study of behaviour. Original research in animal behaviour is published in many scientific journals, but the main ones spanning the whole field are *Animal Behaviour*, *Behaviour* and *Ethology*. The *Journal of Comparative Psychology* also includes a wide variety of relevant articles, mostly in the American tradition of carefully controlled laboratory experiments, while the great growth in functional and evolutionary studies has led to two relatively young journals: *Behavioral Ecology* and *Behavioral Ecology and Sociobiology*. Many brief review articles on topics of current interest are published in the monthly magazine *Trends in Ecology and Evolution*.

Turning to the topics of individual chapters, motor systems, sensory systems and motivation have been relatively neglected at the behavioural level in recent years. An exception is in studies of orientation and navigation: Wehner *et al.* (1996) provide an excellent source of up-to-date articles on this field. By contrast, much work on sensory and motor systems has moved to the neurobiological level (see, for example, Young 1989). The chapters in Halliday & Slater (1983a) provide greater depth than possible here on all three subject areas, while Colgan (1989) gives the most recent comprehensive review of motivation. Of the many books written about animal welfare, that by Dawkins (1980) remains a good introduction.

General issues in behavioural development are dealt with in the book edited by Halliday & Slater (1983c) and, more recently, in many of the chapters in the book edited by Hogan & Bolhuis (1994). On more specific topics, Catchpole & Slater (1995, Chapter 3) give a recent survey of bird song development, Johnston & Morton (1991) discuss parallels between imprinting in birds and face recognition in babies, and Candland (1993) is a fascinating (and salutory) account of what we have learnt from children with impoverished upbringing like Kaspar Hauser and from attempts to train animals to do clever things. There has been a great deal of interest recently in social learning, and whether or not animals are capable of imitation: the reviews in the book edited by Heyes & Galef (1995) provide a useful survey of this field.

As far as the evolution of behaviour is concerned, a recent book dealing with a variety of issues is that edited by Slater & Halliday (1994), while Harvey & Pagel (1991) provide a definitive account of techniques, and pitfalls, in carrying out the comparative method. For most topics in behavioural ecology, one need look no further than the introductory text by Krebs & Davies (1993), now in its third edition, and their more advanced edited volume (Krebs & Davies 1997), now in its fourth. Several of the chapters in earlier editions of the latter also remain worth reading. A lucid recent account of modelling in behavioural ecology, including evolutionarily stable strategies, is the chapter by Harvey in Slater & Halliday (1994). A fine case history of the economic approach, on the topic covered in Box 1.4, is given by Heinrich (1979).

Communication has been a particularly active area of research in the past few years. The book edited by Halliday & Slater (1983b) covered a range of issues in the field, but is now rather dated. The chapter by Johnstone in Krebs & Davies (1997) nicely summarises current points of interest, as does the recent book by Bradbury & Vehrencamp (1998). The book edited by Real (1994) includes chapters by Wiley, discussing topics such as deception and signal detection, and by Ryan, on sexual selection with a section on sensory exploitation. Andersson (1994) is currently the definitive review of sexual selection. Cheney & Seyfarth (1990) provide a fascinating account of their important studies on vervet monkeys.

Many of the chapters in Krebs & Davies (1997) and in Slater & Halliday (1994) also deal with topics discussed here in the chapter on social organisation. In addition there is, of course, a wealth of monographs dealing with the social systems of particular species. That edited by Sherman *et al.* (1991) on naked mole-rats goes in detail into one of the most remarkable systems touched on in Chapter 9, and those by Dunbar (1988) and edited by Smuts *et al.* (1987) survey the great variety of systems found among the primates.

Alcock, J. (1998). *Animal Behavior.* Sixth edition. Sunderland, Mass.: Sinauer.

Andersson, M. (1994). *Sexual Selection.* Princeton, NJ: Princeton University Press.

Bradbury, J.W. & Vehrencamp, S.L. (1998). *Principles of Animal Communication.* Sunderland, Mass.: Sinauer.

Candland, D. K. (1993). *Feral Children and Clever Animals.* Oxford: Oxford University Press.

Catchpole, C. K. & Slater, P. J. B. (1995). *Bird Song. Biological Themes and Variations.* Cambridge: Cambridge University Press.

Cheney, D. L. & Seyfarth, R. L. (1990). *How Monkeys See the World.* Chicago: University of Chicago Press.

Colgan, P. (1989). *Animal Motivation.* London: Chapman & Hall.

Dawkins, M. S. (1980). *Animal Suffering.* London: Chapman & Hall.

(1995). *Unravelling Animal Behaviour.* London: Longman.

Dunbar, R. I. M. (1988). *Primate Social Systems.* London: Croom Helm.

Halliday, T. R. & Slater, P. J. B. (Eds.) (1983a). *Animal Behaviour, Volume 1: Causes and Effects.* Oxford: Blackwell Scientific Publications.

(1983b). *Animal Behaviour, Volume 2: Communication.* Oxford: Blackwell Scientific Publications.

(1983c). *Animal Behaviour, Volume 3: Genes, Development and Learning.* Oxford: Blackwell Scientific Publications.

Harvey, P. H. & Pagel, M. D. (1991). *The Comparative Method in Evolutionary Biology.* Oxford: Oxford University Press.

Heinrich, B. (1979). *Bumblebee Economics.* Cambridge, Mass.: Harvard University Press.

Heyes, C. M. & Galef, B. G. (Eds.) (1996). *Social Learning in Animals: The Roots of Culture.* London: Academic Press.

Hogan, J. A. & Bolhuis, J. J. (Eds.) (1994). *Causal Mechanisms of Behavioural Development.* Cambridge: Cambridge University Press.

Houck, L. D. & Drickamer, L. C. (Eds.) (1996). *Foundations of Animal Behavior.* Chicago: University of Chicago Press.

Johnston, M. H. & Morton, M. (1991). *Biology and Cognitive Development: The Case of Face Recognition.* Oxford: Blackwell Scientific Publications.

Krebs, J. R. & Davies, N. B. (1993). *An Introduction to Behavioural Ecology.* Oxford: Blackwell Scientific Publications.

(Eds.) (1997). *Behavioural Ecology: An Evolutionary Approach.* Oxford: Blackwell Science.

Manning, A. & Dawkins, M. S. (1998). *An Introduction to Animal Behaviour.* Fifth edition. Cambridge: Cambridge University Press.

Real, L. A. (Ed.) (1994). *Behavioral Mechanisms in Evolutionary Ecology.* Chicago: University of Chicago Press.

Sherman, P. W., Jarvis, J. U. M. & Alexander, R. D. (1991). *The Biology of the Naked Mole-Rat.* Princeton, NJ: Princeton University Press.

Slater, P. J. B. & Halliday, T. R. (Eds.) (1994). *Behaviour and Evolution.* Cambridge: Cambridge University Press.

Smuts, B. B., Cheney, D. L., Seyfarth, R. M., Wrangham, R. W. & Struhsaker, P. T. (Eds.) (1987). *Primate Societies.* Chicago: University of Chicago Press.

Wehner, R., Lehrer, M. & Harvey, W. R. (Eds.) (1996). *Navigation. Journal of experimental Biology (Special issue)* **199**(1), 1–261.

Young, D. (1989). *Nerve Cells and Animal Behaviour.* Cambridge: Cambridge University Press.

Glossary

Adaptation Process whereby animals adjust genetically so that a characteristic becomes better adapted to some aspect of the environment. The features so evolved are referred to as adaptations.

Affective signals Signals that convey messages about the internal state of an animal (as opposed to *Referential signals*).

Aggression Term describing a loose categorisation of attack and threat behaviour patterns within a species.

Agonistic behaviour Behaviour patterns associated with fighting and retreat, such as attack, escape, threat, defence and appeasement.

Allogrooming Mutual grooming between animals.

Altricial Those species in which the offspring are relatively helpless at birth and remain for a period in a nest or other hiding place (in contrast to *Precocial*).

Altruism Conferring of a benefit on another individual at a cost to the performer's own fitness. Reciprocal altruism arises where two individuals can recognise each other and take the opposite rôles on different occasions so that both gain in the long run.

Amplexus Posture adopted by spawning frogs and toads with the male clasping the female.

Anisogamy Germ cells (gametes) differing in size between the sexes, as in all higher animals.

Anthropomorphism Ascription of human characteristics to animals.

Antithesis Principle put forward by Darwin that signals with opposite messages also tend to be opposite in form.

Appeasement Behaviour that inhibits attack in situations where animals cannot escape or, as in courtship, where it is disadvantageous for them to do so.

Appetitive behaviour Active searching behaviour shown by animals seeking a goal, contrasted with the consummatory behaviour they show when they reach it.

Arousal Idea that behaviour of animals is influenced by a single internal factor, with a continuum of states between deep sleep and high excitement.

Assortative mating Where individuals choose mates non-randomly in relation to their own characteristics. If positive, the mate chosen is like themselves, if negative, it is unlike themselves.

Attention State in which an animal is more responsive to one aspect of its environment than to others.

Audience effect Influence that the presence of other individuals has on the production of signals, such as alarm calls, by an animal.

Badge Physical feature of an animal that conveys a message about some aspect of its behaviour, such as its fighting ability.

Bait shyness Avoidance by pest animals of poison put out for them.

Behavioural ecology Area of study particularly concerned with the function or selective advantage of behaviour.

Behaviour pattern An action shown by an animal, or group or sequence of such actions that occurs in a predictable combination.

Behaviorism School of psychology founded by J. B. Watson, which rejected introspection and stressed the importance of objective observation.

Brood parasitism Cases where young are raised by individuals that are not their parents, either of a different species or, as with *Egg dumping*, of the same species.

Camouflage Colours or patterns of animals that blend with their background so that predators or prey are less likely to detect them.

Caste Individuals that are adapted to perform a particular task both by their structure and their behaviour, such as workers in ants and bees, and soldiers in termites.

Causation Internal and external factors that are the immediate causes of behaviour and the mechanisms by which these act.

Central place foraging Foraging where the animal involved returns with food to a particular location, such as a nest or hoarding site.

Cheating Behaviour where an individual gains in fitness at the expense of another animal.

Circadian rhythm Physiological or behavioural cycle that is about 24 hours long. Such cycles often diverge from 24 hours in constant conditions, but are normally entrained by the daily light–dark cycle.

Coefficient of relatedness (*r*) Probability that a gene present in one individual is shared by another for reasons of common descent. In diploid species, for identical twins $r = 1.0$, for full siblings 0.5, for half siblings 0.25, etc.

Comfort movements Varied group of activities, including grooming, shaking, stretching and yawning.

Commensalism Association between two species such that one benefits without appreciable cost to the other.

Communication Transmission of a signal from one individual to another such that the sender benefits, on average, from the response of the receiver.

Comparative method Approach to studying behavioural evolution which uses comparisons between species or higher taxonomic groupings.
Competition Situation where there is an insufficient quantity of a resource so that individuals must compete for it and some get less than they need.
Conditioning Learning by association. In classical conditioning two stimuli become associated so that the second comes to elicit a response formerly only elicited by the first. In operant or instrumental conditioning (also known as trial and error learning) the rate of a response is raised or lowered by its association with *Reinforcement.*
Conflict Motivational concept where an animal has an approximately equal tendency to do two things at the same time.
Consciousness Subjective experience of awareness.
Consort relationship Temporary *Pair bond* between a male and a receptive female.
Constraints Factors that limit the capacity of an animal to show the optimal behaviour in a particular situation, ranging from lack of information to the need to satisfy other behavioural demands.
Constraints on learning Expression to describe fact that the learning of some tasks by animals is very much slower or more difficult than that of others.
Consumer demand theory Economic theory, based on the extent to which demand for an item declines as its price increases, which has been applied to studying the needs of animals.
Consummatory act Act that terminates a behavioural sequence and is followed by a switch to another activity.
Contact call Sound that serves to keep individuals in a pair or social group in touch with one another.
Contrafreeloading Phenomenon whereby animals given plenty of food but also a lever that they can press to earn more, will work for some of their food rather than eating only that which is free.
Cooperation Animals assisting one another for mutual benefit.
Courtship Behaviour patterns that precede mating.
Coyness Reluctance to mate, often found in newly formed pairs.
Crypsis See *Camouflage.*
Cultural evolution Changes in the behaviour of animals within a population brought about through social learning rather than through genetic changes.
Cuckoldry Mating by the female of a pair with males other than her partner.
Deception See *Cheating.*
Decision making Process of choosing between different behavioural options.
Deprivation experiment Technique of depriving animals of particular experiences to see how this influences the development of their behaviour.
Desertion Leaving of parental care to one partner or the other at some stage after mating, or the abandoning of offspring by adults after disturbance or in adverse conditions.
Dialect Variation in the form of signals between one place and another, sometimes

restricted to cases where there are sharp boundaries between variants giving a mosaic of signal forms.

Dilution effect When a predator just takes one prey, the larger the group the more the risk to a particular individual is diluted.

Discrimination Ability of an animal to distinguish between two stimuli.

Dishabituation Recovery of a response after *Habituation* following a change in the stimulus.

Disinhibition Idea that *Diplacement activities* arise because two behavioural systems are in *Conflict*, or one system is thwarted, so that the next in priority activity can occur.

Dispersal Movement of animals between their place of birth or hatching and the place where they reproduce, usually most marked in one sex rather than the other as a mechanism of inbreeding avoidance.

Display Behaviour pattern that acts as a signal, the term most often applied to stereotyped visual signals used in courtship and fighting.

Displacement activity Seemingly irrelevant behaviour, such as the grooming or nest building actions sometimes shown by courting or fighting animals.

Distraction display Display shown by a breeding animal whereby it appears to be injured and so leads a potential predator away from its nest.

DNA fingerprinting Technique which uses specific repeated sections in the DNA of individuals to determine their relatedness to each other.

Dominance Situation where one animal repeatedly wins fights with another, or displaces it from a contested resource. In social groups such relationships may be consistent and clear enough for the animals to be placed in a linear dominance hierarchy or peck order.

Drive Psychological term applied to postulated moving forces (e.g. hunger drive, sex drive) underlying the appearance of behaviour.

Duetting Joint calling of two members of a pair, found in gibbons and many tropical birds, with the contributions sometimes so precise and well coordinated that they sound as if they come from a single individual.

Echolocation Orientation mechanism, by producing high pitched sounds and monitoring their reflection off objects, used primarily by bats and some dolphins.

Eclosion Hatching of an adult insect from its pupa.

Egg dumping Laying of eggs in the nests of other members of an animal's own species.

Emancipation Suggested process whereby movements shown during displays have become divorced from their original controlling mechanisms during *Ritualisation*.

Ethogram Inventory, listing and describing all the behaviour patterns shown by a species.

Ethology European school of behaviour study from a biological perspective founded largely by Konrad Lorenz and Niko Tinbergen.

Eusociality Society in which there is cooperation between individuals and division of labour, found primarily in social insects.

Evolutionarily stable strategy (ESS) Strategy which, if adopted by all members of a population, cannot be invaded by a mutant adopting a different strategy.

Exhaustion Way in which animals cease to respond to a stimulus when it is repeatedly presented to them. May be due to muscular fatigue, sensory adaptation or a variety of central processes such as habituation.

Extinction Process whereby learnt behaviour ceases to be performed when no longer appropriate as, for example, when *Reinforcement* is withdrawn.

Extra-pair copulation (EPC) Occurrence of mating by a member of a pair with an individual other than its partner.

Feedback Modification of behaviour in response to its consequences. In negative feedback the consequences suppress the behaviour (e.g. food intake reduces eating), whereas positive feedback enhances it (e.g. sexual stimulation leads to more sexual behaviour).

Fitness Measure of an individual's genetic contribution to the next generation relative to other members of a population. The term inclusive fitness incorporates both the fitness an individual achieves through its own reproductive efforts (direct component) and that it achieves in addition through assisting relatives (indirect component).

Fixed action pattern Term used by ethologists to refer to stereotyped and species typical acts of behaviour.

Fluctuating asymmetry Asymmetry of structure, varying in direction from one individual to another, thought to arise through stresses during development.

Following response Tendency of young precocial birds to follow the first large moving object that they see when they leave the nest, in nature usually the mother.

Function Adaptive significance or selective advantage of a trait or, in other words, how it enhances the fitness of individuals possessing it.

Garcia effect Phenomenon, discovered by John Garcia, that rats given new food eat a small amount at first and only return to eat more later if they have not become ill.

Goal oriented behaviour Behaviour directed towards some goal which, when reached, brings that behaviour to an end.

Graded signal Signal which differs in frequency or intensity depending on the state of the individual showing it.

Group selection Selection proposed as operating at the group level, with groups of animals being selected for or against, rather than at the level of the individual animal.

Habituation Simple form of learning, whereby an animal ceases to respond to a stimulus presented to it repeatedly which is neither noxious nor rewarding.

Handicap principle Idea that many male secondary sexual characteristics are handicaps, the individual advertising his quality by the fact that he has been able to survive despite their costs.

Haplodiploidy Genetic system of some insects, notably ants and bees, whereby males are haploid, developing from unfertilised eggs, whereas females develop from fertilised ones and are thus diploid.

Harem Group of females guarded by a single male who mates with them and drives off other males attempting to do so.

Helper Individual that assists in the rearing of offspring that are not its own, most often an elder sibling of the brood.

Heritability Measure of the extent to which variation in a behaviour pattern or other characteristic is due to genetic rather than environmental causes, and thus likely to be susceptible to selection.

Heterogeneous summation Way in which, while one or two features may be adequate (see *Sign stimulus*), many different characteristics of a stimulus add together to give a response.

Hierarchy See *Dominance.*

Hoarding Storing of food for later use, either in a cache or, as in scatter hoarding, with each item in a separate place.

Home range Area which an animal or group of animals occupies, but which is not necessarily exclusive to that group or defended as in *Territory.*

Homing Ability of animals to return to the same place, usually a breeding area, from a distance, either as part of their normal activities or after displacement in an experiment.

Imitation Copying of the behaviour of one individual by another such that the second acquires a new behaviour pattern.

Imprinting Process whereby young animals learn the characteristics of other individuals early in life and preferentially associate with them thereafter. Filial imprinting leads birds to learn the characteristics of their mother and stay close to her. Sexual imprinting leads them to learn features of the opposite sex which subsequently influence mate choice.

Incentive Characteristic of a stimulus that makes it pleasant or unpleasant to an animal.

Incest Mating of close relatives, usually of siblings or of parents with offspring.

Inclusive fitness See *Fitness.*

Infanticide Killing of dependent young by members of their own species, usually males which thereby gain reproductive access to the infants' mothers.

Innate behaviour Term with variety of possible meanings: behaviour that develops without learning, without practice, without copying from others, or even without any environmental influence. The last was particularly criticised for its implications of inflexibility and genetic determinism, and the term is not now normally used.

Insight Learning that involves the appreciation of complex relationships and may, at its most sophisticated, imply thought and reasoning. The relatively sudden solution to a problem is sometimes attributed to insight.

Instinct Word now fallen from use, which was applied by ethologists to label systems of behaviour, like the *Drives* of psychologists. Instinctive behaviour was sometimes used as synonymous with *Innate behaviour.*

Intention movements Movements shown by an animal just before commencing an activity which indicate to an observer what it is likely to do.

Intelligence Capacity that enables an individual to learn tasks, reason and solve problems.

Intentionality The attributing of intentions to individuals.

Kaspar Hauser Name given to boy found in Nuremberg in 1828 who behaved as a child and later claimed to have been raised by a man in total isolation. *Deprivation experiments* are sometimes called Kaspar Hauser experiments after him.

Kinesis Movement of an animal which is influenced by a stimulus but not oriented in relation to it.

Kin recognition Ability of animals to recognise relatives, either through previous experience with them, or by some system of *Phenotype matching.*

Kin selection Selection that acts both through its direct effects on an individual's own reproductive efforts and through indirect effects favouring altruism towards non-descendent kin.

Kleptoparasitism Habit of some animals of living on food stolen, often violently, from others.

Language Word usually applied only to the verbal communication system of humans, but sometimes also used to refer to similar aspects of animal communication.

Last male advantage Higher probability that eggs will be fertilised by the last male to mate with a female than by earlier males, as described in various species.

Learning Adaptive and relatively permanent change in individual behaviour as a result of past experience which may take many different forms: see *Conditioning, Habituation, Imprinting, Insight.*

Lek Communal display ground, where males set up and defend small *Territories*, packed close together in an arena and not used for feeding or nesting, to which they attract females for mating.

Local enhancement Simple form of social learning in which one animal draws the attention of another to the appropriate stimulus.

Lordosis The posture shown by receptive females of many mammal species in the presence of a male, with hind quarters raised and tail deflected to one side so that the male can mount.

Major histocompatibility complex (MHC) Highly heterozygous part of the genome coding for cell surface glycoproteins. It is responsible for recognition of 'self' by immune systems and may also enable animals to recognise kin by differences in smell between individuals.

Manipulation View that communication in animals consists of one individual (the sender) manipulating the behaviour of another (the receiver) to its own advantage, rather than the sharing of information between them.

Marginal value theorem Hypothesis that an animal foraging optimally should leave a particular patch of food when it has been depleted to the average density of food in the environment as a whole.

Mate guarding Process whereby the male of a mated pair stays close to its partner and stops others from mating with her during her fertile period thereby enhancing his *Paternity certainty.*

Meaning Information gleaned by the recipient of a signal, which can be assessed by an observer from the way in which it responds to that signal (see also *Message*).

Meme Word coined, by analogy with gene, to refer to a unit of cultural transmission.

Message Information that the sender encodes in a signal (see also *Meaning*).

Migration Long distance movements of animals, and especially the regular seasonal movements between different feeding grounds or between breeding and wintering grounds.

Mimicry Resemblance of two animals in behaviour or physical features such that the two are confused to the advantage of one or both of them. In Batesian mimicry an edible mimic resembles a distasteful model. In Müllerian mimicry features are shared by two or more noxious species. In aggressive mimicry a predator resembles a harmless species. The word is also applied to imitation of behaviour, as where birds learn the songs of other species.

Mobbing Response of some animals, especially small birds, to a predator, whereby they approach and fly round calling noisily.

Motivation Internal changes leading an animal to behave differently at different times.

Mounting Position, on the back of the female, adopted by most male birds and mammals during mating.

Mutualism Association of different species to their mutual advantage (see also *Symbiosis*).

Navigation Ability of animals to find their way to a goal regardless of the location from which they start (see also *Homing*).

Neophobia Dislike of novelty, especially where animals reject food that is novel or presented to them in a different place (see also *Bait shyness*).

Nepotism Behaving in such a way as to favour relatives over non-relatives.

Oestrus State in female mammals where they are receptive to a male or 'on heat', usually restricted to around the time of ovulation.

Open field Arena, divided into squares, used to study the activity and exploration of animals.

Operant Term used by Skinnerian psychologists to denote a behaviour pattern that could be modified in some way by its consequences. Thus lever pressing is an operant whose frequency rises when it is followed by delivery of food to a hungry animal.

Optimal foraging Theory, based on the idea that foraging animals are selected to behave in some sense optimally, which tests possible rules that they may be following. The commonest examined is the hypothesis that they maximise the net rate of food intake per unit time spent foraging.

Optimality theory Theory that the particular form of behaviour animals show has been selected because it contributes most to fitness. Studies in this field compare the observed behaviour of animals with the predictions of possible models.

Optimal outbreeding Theory that animals choose mates of an intermediate degree of relatedness to themselves, so avoiding inbreeding and also levels of outbreeding that might result in incompatibility of the genes contributed by the two partners.

Orientation Movement of animals in relation to their environment.

Pair bond Attachment between members of a mated pair causing them to stay together.

Panglossianism Extreme opinion that animals should be perfectly adapted in every possible way. It is named after the character Dr Pangloss in Voltaire's novel Candide, who believed that 'all was for the best in the best of all possible worlds'.

Parental investment Any investment by a parent in an individual offspring that increases that offspring's chances of surviving at a cost to the parent's ability to invest in other offspring.

Paternity uncertainty Situation that males, unless they are able to *Mate guard* continuously, can never be totally certain that their mate's offspring are their own, which may lead them to make a lower investment in caring for the young.

Payoff matrix Matrix showing the payoffs to different strategies in a game theory model.

Peck order See *Dominance.*

Phenotype matching Mechanism of *Kin recognition* whereby an individual compares features of others it encounters with a standard based on its experience of its parents, siblings or self.

Pheromone Chemical passed through the environment from one animal to another, eliciting either a behavioural response or a physiological change in recipients.

Play Difficult category of behaviour to define, because it involves many types of behaviour shown in other contexts, such as fighting or prey capture, but play often lacks the organisation, completeness and 'earnestness' of these. It is almost entirely restricted to mammals, occurs mostly in young ones, and is of uncertain function.

Pleiotropy Phenomenon whereby a single gene may affect a number of different characters.

Polygamy Mating system in which a single individual has two or more mates, either simultaneously or in succession (the latter also called serial monogamy). May be either polygyny, where a male has several females or, more rarely, polyandry, where one female has several males.

Polygyny threshold Point at which polygyny becomes advantageous to a female because her reproductive success is likely to be greater as the second mate of a male controlling good resources than as the only mate of a male with poor ones.

Precocial Species where the young are well developed at birth, and able to move about and feed themselves at an early age (in contrast to *Altricial*).

Predator inspection Phenomenon in fish whereby individuals, sometimes in small groups, leave their shoal and swim rapidly towards a predator before retreating to the shoal once more.

Presentation Receptive posture shown by female monkeys towards a male (see *Lordosis*).

Prisoner's dilemma Game theory model used in the study of cooperation.

Pursuit deterrence Display produced by a prey animal in the presence of a predator, such as stotting in gazelles, that is thought to signal to the predator that it has been seen and would be unlikely to succeed in a chase.

Redundancy Repetition of a signal several times, or inclusion of more detail in it than is necessary for its recognition under ideal conditions, ensuring that it is received and understood.

Referential signals Signals that convey messages about the external world (as opposed to *Affective signals*).

Reflex Automatic and involuntary response that is the simplest form of reaction to external stimuli.

Reinforcement Term used in the study of learning for events that raise or lower the probability of a response, such as food reward (positive reinforcement) or electric shock (negative reinforcement).

Releaser Physical and behavioural features of an animal that have evolved because they elicit a response from other individuals. A subset of *Sign stimuli.*

Resource holding potential (RHP) Capacity of an individual or group to monopolise resources, such as food or mates, and so deny others access to them.

Response Behaviour pattern shown as a result of a stimulus.

Ritualisation Process that occurs during the evolution of displays so that they become more stereotyped and striking, and thus clearly distinguished from the actions from which they evolved.

Satellite male Male that associates with other males that are attractive to females and attempts to intercept approaching females and mate with them.

Scent marking Marking of objects or individuals with scent, either from special glands or in the urine or faeces.

Search image Phenomenon whereby animals that have difficulty spotting cryptic prey, having found one item quickly find others. The suggestion, though controversial, is that they have learnt to see the prey and so have formed a search image of it.

Sensitisation Increase in responsiveness to a stimulus as a result of exposure to it (as opposed to *Habituation*).

Sensitive period Period of time during development when animals exposed to a stimulus are most likely to respond to it, as in song learning or *Imprinting.*

Sensory adaptation Loss of sensitivity in sense organs exposed to a stimulus repeatedly or for a long time.

Sensory exploitation Suggestion that displaying animals may be more effective in eliciting a response if their signals match pre-existing biases in the sensory systems of recipients.

Sign stimulus Key feature of an external stimulus to which an animal responds in a specific way.

Signal Physical form in which a message of communication is transmitted.

Signal detection theory Theory concerned with how signals are detected in a noisy environment.

Skinner box Apparatus devised by B. F. Skinner for the study of *Operant conditioning* in which an animal makes a response, such as bar pressing, for which it receives a *Reinforcement*, such as the delivery of food.

Slave-making Behaviour of certain ants that kidnap the workers (see *Castes*) of other species and use them as workers in their own colony.

Social facilitation Performance of a behaviour pattern by one individual leading others to start behaving in the same way or to show more of that behaviour.

Social learning Influences of social companions on learning, ranging from local enhancement to imitation.

Sociobiology Study of the social behaviour of animals, its ecology and evolution.

Soliciting Posture of a female that acts as an invitation to a male to mate.

Sonagraph Equipment for analysing sounds that produces a plot of frequency against time known as a sonagram.

Song Rather loose term applied to more or less lengthy and complicated sound signals.

Spatial learning Learning about the positions of objects in space.

Spawning Production of eggs and sperm in those animals, such as most fish species, in which fertilisation is external and these are just released into the water.

Specific hunger Appetite for particular nutrients.

Sperm competition Sexual competition after mating, when sperm from different males compete for fertilisation of a female's eggs.

Stereotypies Stereotyped and monotonously repeated movement patterns with no obvious goal or function, especially those shown by caged animals.

Stimulus Internal or external event sensed by an animal and causing it to respond in some way.

Stimulus generalisation Tendency of an animal deprived of the ideal stimulus for a particular behaviour pattern to respond to a wider range of stimuli.

Strategy Mode of behaviour adopted by an individual in a particular situation when it could, either actually or theoretically, behave in a different way.

Stotting See *Pursuit deterrence.*

Stress Physiological state induced in animals by conditions they are unable to tolerate or cope with, such as pain or overcrowding.

Stridulation Method of sound production used by some insects, in which they scrape two specially adapted parts of their body, such as a leg and a wing, together.

Subordinacy Term used to describe the position of animals that lose in fights or defer to other individuals (as opposed to *Dominance*).

Supernormal stimulus Artificial stimulus that is more effective in eliciting a behaviour pattern that the normal stimulus found in nature.

Symbiosis Association between species, normally one in which both benefit (i.e. *Mutualism*).

Synchrony Occurrence of the same behaviour in different individuals at the same time.

Taxis Movement in which an animal orients its body with respect to a stimulus, moving towards or away from it, or taking up a particular angle in relation to it.

Template Suggested sensory or neural recognition system that guides learning, applied especially to song learning in young birds, the template ensuring that they only copy sounds appropriate to their species.

Territory Area occupied by one or more individuals of a species and defended against the intrusions of others (see also *Home Range*).

Threat Behaviour shown in agonistic situations which signals that the animal may attack.

Thwarting Situation where an animal is stopped from performing a behaviour pattern by some external barrier or by the absence of the expected stimulus.

Tit-for-Tat (TFT) Behaviour strategy that may lead to cooperation between individuals that encounter each other repeatedly, in which an individual cooperates on the first encounter and thereafter behaves in each encounter as its companion did in the previous one.

Tradition Behaviour complex shared by individuals in a population and passed from one animal to another by *Social learning*.

Trial and error See *Conditioning*.

Trail following Behaviour of some species, especially of ants, in which workers lay a *Pheromone* trail from sources of food to the nest and others reach the food by following the trail.

Typical intensity Tendency of animal signals to be performed at about the same intensity regardless of the signaller's motivational state. May either make signals unambiguous or hide the signaller's motivation where it is advantageous to do so.

Ultrasound Sounds beyond the limit of human hearing (above about 20,000 cycles per second), used in communication and, especially, *Echolocation*.

Vacuum activity Occurrence of a behaviour pattern in the absence of the external stimulus normally required to elicit it.

Vigilance State of an animal enhancing the likelihood that it will detect an unpredictable event in the environment, such as the appearance of a predator.

List of species names

ant, leaf-cutting	*Acromyrmex lundi*
babbler, Arabian	*Turdoides squamiceps*
baboon, hamadryas	*Papio hamadryas*
baboon, yellow	*Papio cynocephalus*
badger, American	*Taxidea taxus*
badger, European	*Meles meles*
bat, fringe-lipped	*Trachops cirrhosus*
bat, vampire	*Desmodus rotundus*
bear, polar	*Ursus maritimus*
bee, honey	*Apis mellifera*
bee, sweat	*Lasioglossum zephyrum*
bee-eater, red-throated	*Merops bullocki*
bishop, red	*Euplectes oryx*
blackbird, European	*Turdus merula*
boatman, water	*Notonecta glauca*
bonobo	*Pan paniscus*
budgerigar	*Melopsittacus undulatus*
bullfrog	*Rana catesbeana*
bumblebee	*Bombus* sp.
bunting, reed	*Emberiza schoeniclus*
canary	*Serinus canaria*
cat, domestic	*Felis domestica*
chaffinch	*Fringilla coelebs*
chameleon, African	*Chamaeleo chamaeleon*
cheetah	*Acinonyx jubatus*
chicken, domestic	*Gallus domesticus*
chimpanzee	*Pan troglodytes*

colobus, black and white *Colobus guereza*
colobus, red *Colobus badius*
coyote *Canis latrans*
deer, red *Cervus elaphus*
dog, domestic *Canis familiaris*
dog, African wild *Lycaon pictus*
dolphin, bottlenose *Tursiops truncatus*
dove, Barbary *Streptopelia risoria*
duck, mandarin *Aix galericulata*
duck, wood *Aix sponsa*
earthworm *Lumbricus terrestris*
finch, zebra *Taeniopygia guttata*
fish, paradise *Macropodus opercularis*
fish, Siamese fighting *Betta splendens*
flatworm *Dendrocoelum* sp.
fly, tsetse *Glossina morsitans*
flycatcher, pied *Muscicapa hypoleuca*
flycatcher, spotted *Muscicapa striata*
fox, red *Vulpes vulpes*
frog, common *Rana temporaria*
frog, crowned tree *Anotheca spinosa*
frog, grey tree *Hyla versicolor*
frog, green tree *Hyla cinerea*
frog, North American cricket *Acris crepitans*
frog, túngara *Physalaemus pustulosus*
fruitfly *Drosophila* sp.
garganey *Anas querquedula*
gazelle, Thomson's *Gazella thomsoni*
gibbon *Hylobates* sp.
giraffe *Giraffa camelopardalis*
goldeneye *Bucephala clangula*
goldfish *Carassius auratus*
goose, greylag *Anser anser*
gorilla *Gorilla gorilla*
grebe, great-crested *Podiceps cristatus*
gull, black-headed *Larus ridibundus*
gull, Galapagos swallow-tailed *Larus furcatus*
gull, herring *Larus argentatus*
guppy *Poecilia reticulata*
hawk, zone-tailed *Buteo albonotatus*
horse *Equus caballus*
hoverfly *Volucella pellucens*
hyaena, spotted *Hyaena hyaena*

jackal, golden	*Canis aureus*
jay, Florida scrub	*Aphelocoma coerulescens*
kittiwake	*Rissa tridactyla*
leopard	*Panthera pardus*
lion	*Panthera leo*
locust	*Locusta migratoria*
mantis, praying	*Parastagmatoptera unipunctata*
marmoset, common	*Callithris jacchus*
minnow	*Phoxinus phoxinus*
mite, water	*Neumania papillator*
mockingbird, northern	*Mimus polyglottos*
mole-rat, naked	*Heterocephalus glaber*
monkey, Japanese	*Macaca fuscata*
monkey, rhesus	*Macaca mulatta*
monkey, vervet	*Cercopithecus aethiops*
mouse (laboratory)	*Mus musculus*
mouse, spiny	*Acomys cahirinus*
mudskipper	*Boleophthalmus boddarti*
mussel	*Mytilus edulis*
mynah, Indian hill	*Gracula religiosa*
nightjar	*Caprimulgus europaeus*
orangutan	*Pongo pygmaeus*
ostrich	*Struthio camelus*
owl, barn	*Tyto alba*
ox, musk	*Ovibos moschatus*
oystercatcher	*Haematopus ostralegus*
parrot, African grey	*Psittacus erithacus*
peacock	*Pavo cristatus*
pelican	*Pelecanus occidentalis*
pig, domestic	*Sus scrofa*
pigeon (rock dove)	*Columba livia*
pike	*Esox lucius*
pike-cichlid	*Crenicichla alta*
pipit, tree	*Anthus trivialis*
plover, ringed	*Charadrius hiaticula*
prairie dog, black-tailed	*Cynomys ludovicianus*
prinia, tawny-flanked	*Prinia subflava*
rabbit	*Oryctolagus cuniculus*
rat (laboratory)	*Rattus norvegicus*
rattlesnake	*Crotalus adamanteus*
rhinoceros	*Ceratotherium simum*
robin, European	*Erithacus rubecula*
robin-chat, African	*Cossypha heuglini*

salamander, marbled	*Ambystoma opacum*
scorpionfly	*Hylobittacus apicalis*
silkmoth	*Hyalophora cecropia*
snake, garter	*Thamnophis elegans*
sparrow, Harris'	*Zonotrichia querula*
sparrowhawk	*Accipiter nisus*
starling	*Sturnus vulgaris*
stickleback, ten-spined	*Pygosteus pungitius*
stickleback, three-spined	*Gasterosteus aculeatus*
sunfish, bluegill	*Lepomis macrochirus*
swallow, barn	*Hirundo rustica*
tchagra, brown-headed	*Tchagra australis*
tit, blue	*Parus caeruleus*
tit, great	*Parus major*
tit, European long-tailed	*Aegithalos caudatus*
toad, European common	*Bufo bufo*
warbler, garden	*Sylvia borin*
warbler, marsh	*Acrocephalus palustris*
warbler, sedge	*Acrocephalus schoenobaenus*
warbler, willow	*Phylloscopus trochilus*
weaverbird, red-billed	*Quelea quelea*
whale, killer	*Orcinus orca*
wheater, northern	*Oenanthe oenanthe*
wolf	*Canis lupus*
wood-hoopoe, green	*Phoeniculus purpureus*
wildebeest	*Connochaetes taurinus*
zebra, plains	*Equus burchelli*

Index